Riemann–Stieltjes Integral Inequalities for Complex Functions Defined on Unit Circle

With Applications to Unitary Operators in Hilbert Spaces

Riemann–Stieltjes Integral Inequalities for Complex Functions Defined on Unit Circle

With Applications to Unitary Operators in Hilbert Spaces

Silvestru Sever Dragomir

CRC Press
Taylor & Francis Group
Boca Raton London New York

CRC Press is an imprint of the
Taylor & Francis Group, an **informa** business

CRC Press
Taylor & Francis Group
6000 Broken Sound Parkway NW, Suite 300
Boca Raton, FL 33487-2742

Printed on acid-free paper

International Standard Book Number-13: 978-0-367-33710-0 (Hardback)

Visit the Taylor & Francis Web site at
http://www.taylorandfrancis.com

and the CRC Press Web site at
http://www.crcpress.com

To my granddaughters, Audrey and Sienna.

Contents

Preface   ix

CHAPTER   1 ▪ Introduction to Unitary Operators   1

CHAPTER   2 ▪ Ostrowski-Type Inequalities   5

2.1	INTRODUCTION	5
2.2	OSTROWSKI-TYPE INEQUALITIES	7
2.3	A QUADRATURE RULE	21
2.4	APPLICATIONS FOR FUNCTIONS OF UNITARY OPERATORS	25

CHAPTER   3 ▪ Trapezoid-Type Inequalities   35

3.1	INTRODUCTION	35
3.2	TRAPEZOID-TYPE INEQUALITIES	37
3.3	A QUADRATURE RULE	47
3.4	APPLICATIONS FOR UNITARY OPERATORS	50

CHAPTER   4 ▪ Generalized Trapezoid Inequalities   55

4.1	INTRODUCTION	55
4.2	GENERALIZED TRAPEZOID INEQUALITIES	57
4.3	APPLICATIONS FOR FUNCTIONS OF UNITARY OPERATORS	68
4.4	A QUADRATURE RULE	71

CHAPTER 5 ▪ Quasi-Grüss-Type Inequalities 77

5.1 INTRODUCTION 77
5.2 INEQUALITIES FOR RIEMANN–STIELTJES INTEGRAL 79
5.3 APPLICATIONS FOR FUNCTIONS OF UNITARY OPERATORS 90

CHAPTER 6 ▪ Grüss-Type Inequalities 97

6.1 INTRODUCTION 97
6.2 INEQUALITIES FOR RIEMANN–STIELTJES INTEGRAL 99
6.3 APPLICATIONS FOR FUNCTIONS OF UNITARY OPERATORS 109

CHAPTER 7 ▪ Inequalities for Bounded Functions 115

7.1 SOME IDENTITIES 115
7.2 INEQUALITIES FOR BOUNDED FUNCTIONS 121
7.3 QUASI-GRÜSS-TYPE INEQUALITIES 126
7.4 GRÜSS-TYPE INEQUALITIES 134
7.5 INEQUALITIES FOR UNITARY OPERATORS 138

Bibliography 141

Preface

Linear Operator Theory in Hilbert spaces plays a central role in contemporary mathematics with numerous applications for Partial Differential Equations, in Approximation Theory, Optimization Theory, Numerical Analysis, Probability Theory & Statistics and other fields.

The main aim of this book is to present several results related to functions of unitary operators on complex Hilbert spaces obtained by the author in a sequence of recent research papers. The fundamental tools to obtain these results are provided by some new Riemann–Stieltjes integral inequalities of continuous integrands on the complex unit circle and integrators of bounded variation.

The monograph starts with a short introductory chapter where some fundamental facts related to the spectral representation of functions of unitary operators in terms of the *spectral family* of a unitary operator are presented.

In the second chapter, we present several Ostrowski-type inequalities for the Riemann–Stieltjes integral of continuous integrands on the complex unit circle and integrators of bounded variation. Applications for Lipschitzian functions of unitary operators are also given.

In the third chapter, we present the corresponding trapezoid-type results, while the generalized trapezoid inequalities are considered in Chapter 4. All these inequalities for the Riemann–Stieltjes integral are then applied for functions of unitary operators, with some particular examples provided.

In the fifth chapter, we present some quasi-Grüss-type inequalities for the Riemann–Stieltjes integral of a continuous function defined on the complex unit circle and an integrator of bounded variation that are thereafter applied for approximating continuous functions of unitary operators. The more general of two continuous integrands on the complex unit circle and integrators of bounded variation with application to the approximation of a product of two functions of unitary operator are provided in the next chapter.

The seventh chapter contains various Riemann–Stieltjes integral identities with only some of them applied to obtain certain Grüss- and quasi-Grüss-type inequalities for a class of bounded functions on the complex unit circle. This chapter can also be used by the interested reader to explore several other inequalities that are not explicitly mentioned in the monograph.

For the sake of completeness, all the results presented are completely proved and the original references from which they were first obtained are mentioned.

The book is intended for use by researchers in various fields of Linear Operator Theory and Mathematical Inequalities, domains which have grown exponentially in the last decade, as well as by postgraduate students and scientists applying inequalities in their specific areas.

CHAPTER 1

Introduction to Unitary Operators

W**E** SAY THAT the bounded linear operator $U :$ $H \to H$ on the Hilbert space H is *unitary* iff $U^* = U^{-1}$.

It is well known that (see for instance [27, p. 275-p. 276]), if U is a unitary operator, then there exists a family of *projections* $\{E_\lambda\}_{\lambda \in [0, 2\pi]}$, called the *spectral family* of U with the following properties:

a) $E_\lambda \leq E_\mu$ for $0 \leq \lambda \leq \mu \leq 2\pi$;

b) $E_0 = 0$ and $E_{2\pi} = 1_H$ (the *identity operator on H*);

c) $E_{\lambda+0} = E_\lambda$ for $0 \leq \lambda < 2\pi$;

d) $U = \int_0^{2\pi} e^{i\lambda} dE_\lambda$, where the integral is of Riemann–Stieltjes type.

Moreover, if $\{F_\lambda\}_{\lambda \in [0, 2\pi]}$ is a family of projections satisfying the requirements a)-d) above for the operator U, then $F_\lambda = E_\lambda$ for all $\lambda \in [0, 2\pi]$.

Also, for every continuous complex-valued function $f : \mathcal{C}(0, 1) \to \mathbb{C}$ on the complex unit circle $\mathcal{C}(0, 1)$, we have

$$f(U) = \int_0^{2\pi} f\left(e^{i\lambda}\right) dE_\lambda \qquad (1.1)$$

where the integral is taken in the Riemann–Stieltjes sense.

In particular, we have the equalities

$$\langle f(U)x, y \rangle = \int_0^{2\pi} f\left(e^{i\lambda}\right) d\langle E_\lambda x, y \rangle \qquad (1.2)$$

and

$$\|f(U)x\|^2 = \int_0^{2\pi} \left| f\left(e^{i\lambda}\right) \right|^2 d\|E_\lambda x\|^2$$
$$= \int_0^{2\pi} \left| f\left(e^{i\lambda}\right) \right|^2 d\langle E_\lambda x, x \rangle, \qquad (1.3)$$

for any x, $y \in H$.

Examples of such functions of unitary operators are

$$\exp(U) = \int_0^{2\pi} \exp\left(e^{i\lambda}\right) dE_\lambda$$

and

$$U^n = \int_0^{2\pi} e^{in\lambda} dE_\lambda$$

for n, an integer.

We can also define the *trigonometric functions* for a unitary operator U by

$$\sin(U) = \int_0^{2\pi} \sin\left(e^{i\lambda}\right) dE_\lambda \text{ and } \cos(U) = \int_0^{2\pi} \cos\left(e^{i\lambda}\right) dE_\lambda$$

and the *hyperbolic functions* by

$$\sinh(U) = \int_0^{2\pi} \sinh\left(e^{i\lambda}\right) dE_\lambda \text{ and } \cosh(U)$$
$$= \int_0^{2\pi} \cosh\left(e^{i\lambda}\right) dE_\lambda$$

where

$$\sinh(z) := \frac{1}{2}\left[\exp z - \exp(-z)\right] \text{ and}$$
$$\cosh(z) := \frac{1}{2}\left[\exp z + \exp(-z)\right], z \in \mathbb{C}.$$

We can prove the following result that provides an upper bound for the total variation of the function $\mathbb{R} \ni \lambda \mapsto \langle E_\lambda x, y \rangle \in \mathbb{C}$ on an interval $[\alpha, \beta]$:

Theorem 1.1 *Let* $\{E_\lambda\}_{\lambda\in[0,2\pi]}$ *be the spectral family of the unitary operator U. Then for any* $x, y \in H$ *and* $\alpha < \beta$ *with* $[\alpha, \beta] \subset [0, 2\pi]$ *we have the inequality*

$$\left[\bigvee_{\alpha}^{\beta}\left(\left\langle E_{(\cdot)}x, y\right\rangle\right)\right]^2 \leq \left\langle (E_\beta - E_\alpha)\, x, x\right\rangle \left\langle (E_\beta - E_\alpha)\, y, y\right\rangle,$$

(1.4)

where $\bigvee_{\alpha}^{\beta}\left(\left\langle E_{(\cdot)}x, y\right\rangle\right)$ *denotes the total variation of the function* $\left\langle E_{(\cdot)}x, y\right\rangle$ *on* $[\alpha, \beta]$.

Proof. Now, if $d : \alpha = t_0 < t_1 < \ldots < t_{n-1} < t_n = \beta$ is an arbitrary partition of the interval $[\alpha, \beta]$, then we have by Schwarz's inequality for positive operators that

$$\bigvee_{\alpha}^{\beta}\left(\left\langle F_{(\cdot)}x, y\right\rangle\right)$$

(1.5)

$$= \sup_{d}\left\{\sum_{i=0}^{n-1}\left|\left\langle (E_{t_{i+1}} - E_{t_i})\, x, y\right\rangle\right|\right\}$$

$$\leq \sup_{d}\left\{\sum_{i=0}^{n-1}\left[\left\langle (E_{t_{i+1}} - E_{t_i})\, x, x\right\rangle^{1/2}\left\langle (E_{t_{i+1}} - E_{t_i})\, y, y\right\rangle^{1/2}\right]\right\}$$

$$:= B.$$

By the Cauchy-Bunyakovsky-Schwarz inequality for sequences of real numbers, we also have that

$$\sum_{i=0}^{n-1}\left[\left\langle (E_{t_{i+1}} - E_{t_i})\, x, x\right\rangle^{1/2}\left\langle (E_{t_{i+1}} - E_{t_i})\, y, y\right\rangle^{1/2}\right]$$

(1.6)

$$\leq \left[\sum_{i=0}^{n-1}\left\langle (E_{t_{i+1}} - E_{t_i})\, x, x\right\rangle\right]^{1/2}\left[\sum_{i=0}^{n-1}\left\langle (E_{t_{i+1}} - E_{t_i})\, y, y\right\rangle\right]^{1/2}$$

$$= \left[\left\langle (E_\beta - E_\alpha)\, x, x\right\rangle\right]^{1/2}\left[\left\langle (E_\beta - E_\alpha)\, y, y\right\rangle\right]^{1/2}$$

for any $x, y \in H$. Taking the supremum over d in (1.6) we get

$$B \leq \left[\left\langle (E_\beta - E_\alpha)\, x, x\right\rangle\right]^{1/2}\left[\left\langle (E_\beta - E_\alpha)\, y, y\right\rangle\right]^{1/2}$$

for any $x, y \in H$, which together with (1.5) produce the desired result (1.4).

Remark 1 *For $\alpha = 0$ and $\beta = 2\pi$ we get from (1.4) the inequality*

$$\bigvee_0^{2\pi} \left(\left\langle E_{(\cdot)}x, y \right\rangle \right) \leq \|x\| \, \|y\| \tag{1.7}$$

for any $x, y \in H$.

CHAPTER 2

Ostrowski-Type Inequalities

IN THIS CHAPTER we present some Ostrowski-type Riemann–Stieltjes integral inequalities for continuous complex-valued integrands and various classes of bounded variation integrators. Natural applications for functions of unitary operators in Hilbert spaces are provided as well.

2.1 INTRODUCTION

The problem of approximating the *Riemann–Stieltjes integral* $\int_a^b f(t) \, du(t)$ by the quantity $f(x) [u(b) - u(a)]$, which is a natural generalization of the Ostrowski problem analyzed in 1937 (see [28]), was apparently first considered in the literature by the author in 2000 (see [8]) where he obtained the following result:

$$\left| \left[u\left(b\right) - u\left(a\right) \right] f\left(x\right) - \int_a^b f\left(t\right) du\left(t\right) \right| \tag{2.1}$$

$$\leq H \left[\left(x - a\right)^r \bigvee_a^x \left(f\right) + \left(b - x\right)^r \bigvee_x^b \left(f\right) \right]$$

$$\leq H \times \begin{cases} \left[\left(x-a\right)^r + \left(b-x\right)^r\right] \left[\frac{1}{2} \bigvee_a^b \left(f\right) + \frac{1}{2}\left| \bigvee_a^x \left(f\right) - \bigvee_x^b \left(f\right) \right|\right]; \\ \left[\left(x-a\right)^{qr} + \left(b-x\right)^{qr}\right]^{\frac{1}{q}} \left[\left(\bigvee_a^x \left(f\right)\right)^p + \left(\bigvee_x^b \left(f\right)\right)^p\right]^{\frac{1}{p}} \\ \qquad\qquad\qquad \text{if} \quad p > 1, \; \frac{1}{p} + \frac{1}{q} = 1; \\ \left[\frac{1}{2}\left(b-a\right) + \left|x - \frac{a+b}{2}\right|\right]^r \bigvee_a^b \left(f\right); \end{cases}$$

for each $x \in [a, b]$, provided f is of *bounded variation* on $[a, b]$, $\bigvee_a^b \left(f\right)$ is its *total variation* on $[a, b]$, while $u : [a, b] \to \mathbb{R}$ is $r - H$-*Hölder continuous*, i.e., we recall that:

$$\left| u\left(x\right) - u\left(y\right) \right| \leq H \left| x - y \right|^r \quad \text{for each } x, y \in [a, b]. \tag{2.2}$$

The dual case, i.e., when the *integrand* f is $q - K$-Hölder continuous and the *integrator* u is of bounded variation was obtained by the author in 2001 and can be stated as [9]

$$\left| \left[u\left(b\right) - u\left(a\right) \right] f\left(x\right) - \int_a^b f\left(t\right) du\left(t\right) \right| \tag{2.3}$$

$$\leq K \left[\frac{1}{2}\left(b - a\right) + \left| x - \frac{a+b}{2} \right| \right]^q \bigvee_a^b \left(u\right)$$

for each $x \in [a, b]$.

The above inequalities provide, as important consequences, the following *midpoint inequalities*:

$$\left| \left[u\left(b\right) - u\left(a\right) \right] f\left(\frac{a+b}{2}\right) - \int_a^b f\left(t\right) du\left(t\right) \right| \tag{2.4}$$

$$\leq \begin{cases} \frac{1}{2^r}\left(b-a\right)^r H \bigvee_a^b \left(f\right) \\ \frac{1}{2^q}\left(b-a\right)^q K \bigvee_a^b \left(u\right), \end{cases}$$

which can be numerically implemented and provide a quadrature rule for approximating the Riemann–Stieltjes integral $\int_a^b f(t)\, du(t)$.

Motivated by the above results, we provide in the current chapter upper bounds for the magnitude of the difference

$$f\left(e^{is}\right)[u(b) - u(a)] - \int_a^b f\left(e^{it}\right) du(t) \text{ with } s \in [a,b] \subseteq [0, 2\pi]$$

for continuous complex-valued function $f : \mathcal{C}(0,1) \to \mathbb{C}$ defined on the complex unit circle $\mathcal{C}(0,1)$ and various subclasses of functions $u : [a,b] \subseteq [0,2\pi] \to \mathbb{C}$ of bounded variation. Natural applications for functions of unitary operators in Hilbert spaces are provided as well.

2.2   OSTROWSKI-TYPE INEQUALITIES

Theorem 2.1 (Dragomir 2015, [19]) *Assume that* $f : \mathcal{C}(0,1) \to \mathbb{C}$ *satisfies the following Hölder-type condition*

$$|f(z) - f(w)| \le H |z - w|^r \tag{2.5}$$

for any $w, z \in \mathcal{C}(0,1)$, *where* $H > 0$ *and* $r \in (0,1]$ *are given.*

If $[a,b] \subseteq [0, 2\pi]$ *and the function* $u : [a,b] \to \mathbb{C}$ *is of bounded variation on* $[a,b]$, *then*

$$\left| f\left(e^{is}\right)[u(b) - u(a)] - \int_a^b f\left(e^{it}\right) du(t) \right| \tag{2.6}$$

$$\le 2^r H \max_{t \in [a,b]} \left| \sin\left(\frac{s-t}{2}\right) \right|^r \bigvee_a^b (u)$$

for any $s \in [a,b]$.

Proof. Observe that

$$f\left(e^{is}\right)[u(b) - u(a)] - \int_a^b f\left(e^{it}\right) du(t)$$

$$= \int_a^b \left[f\left(e^{is}\right) - f\left(e^{it}\right) \right] du(t) \tag{2.7}$$

for any $s \in [a, b]$.

It is known that if $p : [c, d] \to \mathbb{C}$ is a continuous function and $v : [c, d] \to \mathbb{C}$ is of bounded variation, then the Riemann–Stieltjes integral $\int_c^d p(t)\, dv(t)$ exists and the following inequality holds [1]

$$\left| \int_c^d p(t)\, dv(t) \right| \leq \max_{t \in [c,d]} |p(t)| \bigvee_c^d (v). \qquad (2.8)$$

Applying the property (2.8) to the identity (2.7) and utilizing the Hölder-type condition (2.5) we have successively

$$\left| f\left(e^{is}\right) [u(b) - u(a)] - \int_a^b f\left(e^{it}\right) du(t) \right| \qquad (2.9)$$

$$= \max_{t \in [a,b]} \left| f\left(e^{is}\right) - f\left(e^{it}\right) \right| \bigvee_a^b (u) \leq H \max_{t \in [a,b]} \left| e^{is} - e^{it} \right|^r \bigvee_a^b (u).$$

Since

$$\left| e^{is} - e^{it} \right|^2 = \left| e^{is} \right|^2 - 2 \operatorname{Re} \left(e^{i(s-t)} \right) + \left| e^{it} \right|^2$$

$$= 2 - 2\cos(s - t) = 4\sin^2 \left(\frac{s-t}{2} \right)$$

for any $t, s \in \mathbb{R}$, then

$$\left| e^{is} - e^{it} \right|^r = 2^r \left| \sin \left(\frac{s-t}{2} \right) \right|^r \qquad (2.10)$$

for any $t, s \in \mathbb{R}$.

Now, by (2.9) and (2.10) we deduce the desired result (2.6).

Remark 2 *If $a = 0$ and $b = 2\pi$, then for any $s \in [0, 2\pi]$ there exists a unique $t \in [0, 2\pi]$ such that $\frac{1}{2}|t - s| = \frac{\pi}{2}$, therefore $\max_{t \in [0,2\pi]} \left| \sin \left(\frac{s-t}{2} \right) \right| = 1$ for all $s \in [0, 2\pi]$ and we deduce from (2.6) the following inequality of interest*

$$\left| f\left(e^{is}\right) [u(2\pi) - u(0)] - \int_0^{2\pi} f\left(e^{it}\right) du(t) \right| \leq 2^r H \bigvee_0^{2\pi} (u)$$

$$(2.11)$$

that holds for each $s \in [0, 2\pi]$.

Remark 3 *If $[a, b] \subset [0, 2\pi]$ and $0 < b - a \leq \pi$, then for all $t, s \in [a, b]$ we have $\frac{1}{2} |t - s| \leq \frac{1}{2} (b - a) \leq \frac{\pi}{2}$. Since the function $\sin$ is increasing on $\left[0, \frac{\pi}{2}\right]$, then we have successively that*

$$\max_{t \in [a,b]} \left| \sin \left(\frac{s - t}{2} \right) \right| = \sin \left(\max_{t \in [a,b]} \frac{1}{2} |t - s| \right) \qquad (2.12)$$

$$= \sin \left(\frac{1}{2} \max \{ b - s, s - a \} \right)$$

$$= \sin \left(\frac{1}{4} (b - a) + \frac{1}{2} \left| s - \frac{a + b}{2} \right| \right)$$

for any $s \in [a, b]$.

Therefore, under the assumptions of Theorem 2.1 and if $[a, b] \subset [0, 2\pi]$ with $0 < b - a \leq \pi$, then

$$\left| f \left(e^{is} \right) [u (b) - u (a)] - \int_a^b f \left(e^{it} \right) du (t) \right| \qquad (2.13)$$

$$\leq 2^r H \sin^r \left[\frac{1}{4} (b - a) + \frac{1}{2} \left| s - \frac{a + b}{2} \right| \right] \bigvee_a^b (u)$$

$$\leq 2^r H \sin^r \left[\frac{1}{2} (b - a) \right] \bigvee_a^b (u)$$

for all $s \in [a, b]$.

In particular, the best inequality we can get from (2.13) is incorporated in

$$\left| f \left(e^{\frac{a+b}{2} i} \right) [u (b) - u (a)] - \int_a^b f \left(e^{it} \right) du (t) \right| \qquad (2.14)$$

$$\leq 2^r H \sin^r \left[\frac{1}{4} (b - a) \right] \bigvee_a^b (u).$$

The case when $f : \mathcal{C} (0, 1) \to \mathbb{C}$ satisfies the Lipschitz condition $|f (z) - f (w)| \leq L |z - w|$ for any $w, z \in \mathcal{C} (0, 1)$, where $L > 0$ is given, is of interest due to various examples one can consider. Also in this case we can show that the corresponding version of the inequality (2.15) is sharp.

Corollary 2.2 (Dragomir 2015, [19]) *Assume that* f : $\mathcal{C}(0,1) \to \mathbb{C}$ *is Lipschitzian with the constant* $L > 0$ *on the circle* $\mathcal{C}(0,1)$. *If* $[a,b] \subset [0,2\pi]$ *with* $0 < b - a \leq \pi$ *and the function* $u : [a,b] \to \mathbb{C}$ *is of bounded variation on* $[a,b]$, *then we have*

$$\left| f\left(e^{\frac{a+b}{2}i}\right)[u(b) - u(a)] - \int_a^b f\left(e^{it}\right) du(t) \right|$$

$$\leq 2L \sin\left[\frac{1}{4}(b-a)\right] \bigvee_a^b (u). \quad (2.15)$$

The constant 2 *cannot be replaced by a smaller quantity.*

Proof. We need to prove only the sharpness of the constant 2.

If we consider the function $f : \mathbb{C} \to \mathbb{C}$, $f(z) = z$, then obviously f is Lipschitzian with the constant $L = 1$. Also, consider in (2.15) $a = 0$ and $b = \pi$ to get

$$\left| i[u(\pi) - u(0)] - \int_0^\pi e^{it} du(t) \right| \leq \sqrt{2} \bigvee_0^\pi (u). \quad (2.16)$$

Utilizing the *integration by parts formula* for the Riemann–Stieltjes integral we have

$$\int_0^\pi e^{it} du(t) = e^{it} u(t)\Big|_0^\pi - i \int_0^\pi e^{it} u(t) dt$$

$$= -u(\pi) - u(0) - i \int_0^\pi e^{it} u(t) dt$$

and replacing into the inequality (2.16) we deduce

$$\left| i[u(\pi) - u(0)] + u(\pi) + u(0) + i \int_0^\pi e^{it} u(t) dt \right| \leq \sqrt{2} \bigvee_0^\pi (u)$$

which is equivalent with

$$\left| (i-1) u(\pi) + (i+1) u(0) - \int_0^\pi e^{it} u(t) dt \right| \leq \sqrt{2} \bigvee_0^\pi (u)$$

$$(2.17)$$

that holds for any functions of bounded variation $u : [0, \pi] \to \mathbb{C}$ and is of interest in itself.

Now, assume that there exists a constant $C > 0$ such that

$$\left| (i - 1) u (\pi) + (i + 1) u (0) - \int_0^\pi e^{it} u (t) \, dt \right| \leq C \bigvee_0^\pi (u)$$

(2.18)

for any functions of bounded variation $u : [0, \pi] \to \mathbb{C}$.

Consider the function $u : [0, \pi] \to \mathbb{R}$ with

$$u (t) := \begin{cases} 0 \text{ if } 0 \leq t < \pi \\ \\ 1 \text{ if } t = \pi. \end{cases}$$

Then u is of bounded variation, $\int_0^\pi e^{it} u (t) \, dt = 0, \bigvee_0^\pi (u) = 1$ and from (2.18) we get $C \geq \sqrt{2}$ showing that (2.18) is sharp and therefore (2.15) is sharp.

Remark 4 *The case of the Riemann integral, namely when $u (t) = t, t \in [a, b] \subseteq [0, 2\pi]$, is as follows*

$$\left| f \left(e^{is} \right) - \frac{1}{b - a} \int_a^b f \left(e^{it} \right) dt \right| \leq 2^r H \max_{t \in [a,b]} \left| \sin \left(\frac{s - t}{2} \right) \right|^r$$

(2.19)

for any $s \in [a, b]$ provided that $f : \mathcal{C} (0, 1) \to \mathbb{C}$ satisfies the Hölder-type condition (2.5).

When u is an integral, then the following weighted integral inequality also holds.

Remark 5 *If $w : [a, b] \subseteq [0, 2\pi] \to \mathbb{C}$ is Lebesgue integrable on $[a, b]$ and $f : \mathcal{C} (0, 1) \to \mathbb{C}$ satisfies the Hölder-type condition (2.5), then*

$$\left| f \left(e^{is} \right) \int_a^b w (t) \, dt - \int_a^b f \left(e^{it} \right) w (t) \, dt \right|$$

(2.20)

$$\leq 2^r H \max_{t \in [a,b]} \left| \sin \left(\frac{s - t}{2} \right) \right|^r \int_a^b |w (t)| \, dt$$

for any $s \in [a, b]$.

In particular, if $w(t) \geq 0$ for $t \in [a, b]$ and $\int_a^b w(t)\, dt > 0$ then

$$\left| f\left(e^{is}\right) - \frac{1}{\int_a^b w(t)\, dt} \int_a^b f\left(e^{it}\right) w(t)\, dt \right| \qquad (2.21)$$

$$\leq 2^r H \max_{t \in [a,b]} \left| \sin\left(\frac{s-t}{2}\right) \right|^r$$

for any $s \in [a, b]$.

Theorem 2.3 (Dragomir 2015, [19]) *Assume that* $f : \mathcal{C}(0,1) \to \mathbb{C}$ *is Lipschitzian with the constant* $L > 0$ *on the circle* $\mathcal{C}(0,1)$. *If* $[a,b] \subseteq [0, 2\pi]$ *and the function* $u : [a,b] \to \mathbb{C}$ *is Lipschitzian with the constant* $K > 0$ *on* $[a,b]$, *then*

$$\left| f\left(e^{is}\right) [u(b) - u(a)] - \int_a^b f\left(e^{it}\right) du(t) \right| \qquad (2.22)$$

$$\leq 4LK \left[\sin^2\left(\frac{s-a}{4}\right) + \sin^2\left(\frac{b-s}{4}\right) \right] \leq 8LK \sin^2\left(\frac{b-a}{4}\right)$$

for any $s \in [a, b]$.

Proof. It is well known that if $p : [a,b] \to \mathbb{C}$ is a Riemann integrable function and $v : [a,b] \to \mathbb{C}$ is M-Lipschitzian, then the Riemann–Stieltjes integral $\int_a^b p(t)\, dv(t)$ exists and the following inequality holds

$$\left| \int_a^b p(t)\, dv(t) \right| \leq M \int_a^b |p(t)|\, dt. \qquad (2.23)$$

Utilizing the property (2.23), we have from (2.7) that

$$\left| f\left(e^{is}\right) [u(b) - u(a)] - \int_a^b f\left(e^{it}\right) du(t) \right| \qquad (2.24)$$

$$= \left| \int_a^b \left[f\left(e^{is}\right) - f\left(e^{it}\right) \right] du(t) \right|$$

$$\leq K \int_a^b \left| f\left(e^{is}\right) - f\left(e^{it}\right) \right| dt \leq KL \int_a^b \left| e^{is} - e^{it} \right| dt$$

for any $s \in [a, b]$.

Since, by (2.10), $\left| e^{is} - e^{it} \right| = 2 \left| \sin \left(\frac{s-t}{2} \right) \right|$ for any $t, s \in \mathbb{R}$, then

$$\int_a^b \left| e^{is} - e^{it} \right| dt = 2 \int_a^b \left| \sin \left(\frac{s-t}{2} \right) \right| dt \qquad (2.25)$$

$$= 2 \left[\int_a^s \sin \left(\frac{s-t}{2} \right) dt + \int_s^b \sin \left(\frac{t-s}{2} \right) dt \right]$$

$$= 2 \left[1 - \cos \left(\frac{s-a}{2} \right) \right] + 2 \left[1 - \cos \left(\frac{b-s}{2} \right) \right]$$

$$= 4 \left[\sin^2 \left(\frac{s-a}{4} \right) + \sin^2 \left(\frac{b-s}{4} \right) \right]$$

$$\leq 8 \sin^2 \left(\frac{b-a}{4} \right)$$

for any $s \in [a, b] \subseteq [0, 2\pi]$, and the inequality (2.22) is proved.

The best inequality we can get from (2.22) is incorporated in:

Corollary 2.4 (Dragomir 2015, [19]) *With the assumptions in Theorem 2.3 we have the inequality*

$$\left| f \left(e^{\frac{a+b}{2} i} \right) [u(b) - u(a)] - \int_a^b f \left(e^{it} \right) du(t) \right|$$

$$\leq 8LK \sin^2 \left(\frac{b-a}{8} \right). \qquad (2.26)$$

The multiplicative constant 8 cannot be replaced by a smaller quantity.

Proof. We need to prove only the sharpness of the constant.

If we consider the function $f : \mathbb{C} \to \mathbb{C}$, $f(z) = z$, then obviously f is Lipschitzian with the constant $L = 1$. Also, consider in (2.26) $a = 0$ and $b = 2\pi$ to get

$$\left| - [u(2\pi) - u(0)] - \int_0^{2\pi} e^{it} du(t) \right| \leq 4K. \qquad (2.27)$$

Utilizing the integration by parts formula for the Riemann–Stieltjes integral, we have

$$\int_0^{2\pi} e^{it} du\,(t) = e^{it} u\,(t)\Big|_0^{2\pi} - i \int_0^{2\pi} e^{it} u\,(t)\,dt$$

$$= u\,(2\pi) - u\,(0) - i \int_0^{2\pi} e^{it} u\,(t)\,dt.$$

which inserted in (2.27) produces the inequality

$$\left| -2\left[u\,(2\pi) - u\,(0)\right] + i \int_0^{2\pi} e^{it} u\,(t)\,dt \right| \le 4K$$

which is equivalent with

$$\left| \int_0^{2\pi} e^{it} u\,(t)\,dt - \frac{2}{i}\left[u\,(2\pi) - u\,(0)\right] \right| \le 4K \qquad (2.28)$$

that holds for any K-Lipschitzian function $u : [0, 2\pi] \to \mathbb{C}$ and is of interest in itself.

Now, assume that the inequality (2.28) holds with a constant $D > 0$, namely

$$\left| \int_0^{2\pi} e^{it} u\,(t)\,dt - \frac{2}{i}\left[u\,(2\pi) - u\,(0)\right] \right| \le DK \qquad (2.29)$$

for any K-Lipschitzian function $u : [0, 2\pi] \to \mathbb{C}$.

Consider $u : [0, 2\pi] \to \mathbb{R}$, $u\,(t) = |t - \pi|$. Then, by the continuity property of the modulus we have that u is Lipschitzian with the constant $K = 1$.

We also have that

$$\int_0^{2\pi} e^{it} u\,(t)\,dt = \int_0^{2\pi} e^{it} |t - \pi|\,dt$$

$$= \int_0^{2\pi} |t - \pi|\,(\cos t + i \sin t)\,dt$$

$$= \int_0^{2\pi} |t - \pi| \cos t\,dt + i \int_0^{2\pi} |t - \pi| \sin t\,dt.$$

Observe that, by symmetry reasons, $\int_0^{2\pi} |t - \pi| \sin t \, dt = 0$ and

$$
\int_0^{2\pi} |t - \pi| \cos t \, dt = 2 \int_0^{\pi} (\pi - t) \cos t \, dt
$$
$$
= 2 \left[(\pi - t) \sin t |_0^{\pi} + \int_0^{\pi} \sin t \, dt \right] = 4
$$

and by (2.29) we get $D \geq 4$ which proves the desired sharpness of the constant 8 in (2.26).

Remark 6 *If $u(t) = t, t \in [a, b]$, then we get from (2.22) and (2.26) the following inequalities for the Riemann integral*

$$
\left| f\left(e^{is}\right)(b - a) - \int_a^b f\left(e^{it}\right) dt \right| \tag{2.30}
$$
$$
\leq 4L \left[\sin^2 \left(\frac{s - a}{4} \right) + \sin^2 \left(\frac{b - s}{4} \right) \right] \leq 8L \sin^2 \left(\frac{b - a}{4} \right)
$$

for any $s \in [a, b]$ and

$$
\left| f\left(e^{\frac{a+b}{2}i}\right)(b - a) - \int_a^b f\left(e^{it}\right) dt \right| \leq 8L \sin^2 \left(\frac{b - a}{8} \right),
\tag{2.31}
$$

provided that $f : \mathcal{C}(0, 1) \to \mathbb{C}$ is Lipschitzian with the constant $L > 0$ on the circle $\mathcal{C}(0, 1)$.

Remark 7 *If $w : [a, b] \subseteq [0, 2\pi] \to \mathbb{C}$ is essentially bounded on $[a, b]$ and $f : \mathcal{C}(0, 1) \to \mathbb{C}$ is Lipschitzian with the constant $L > 0$ on the circle $\mathcal{C}(0, 1)$, then we have the following weighted integral inequality*

$$
\left| f\left(e^{is}\right) \int_a^b w(t) \, dt - \int_a^b f\left(e^{it}\right) w(t) \, dt \right| \tag{2.32}
$$
$$
\leq 4L \|w\|_\infty \left[\sin^2 \left(\frac{s - a}{4} \right) + \sin^2 \left(\frac{b - s}{4} \right) \right]
$$
$$
\leq 8L \|w\|_\infty \sin^2 \left(\frac{b - a}{4} \right)
$$

for any $s \in [a, b]$ where $\|w\|_\infty := \operatorname{ess\,sup}_{t \in [a,b]} |w(t)|$.

In particular, we have

$$\left| f\left(e^{\frac{a+b}{2}i}\right) \int_a^b w\left(t\right) dt - \int_a^b f\left(e^{it}\right) w\left(t\right) dt \right| \qquad (2.33)$$

$$\leq 8L \left\|w\right\|_\infty \sin^2\left(\frac{b-a}{8}\right).$$

The case of monotonic integrators is as follows:

Theorem 2.5 (Dragomir 2015, [19]) *Assume that* f : $\mathcal{C}\left(0,1\right) \to \mathbb{C}$ *is Lipschitzian with the constant* $L > 0$ *on the circle* $\mathcal{C}\left(0,1\right).$ *If* $[a,b] \subseteq [0,2\pi]$ *and the function* $u : [a,b] \to \mathbb{R}$ *is monotonic nondecreasing on* $[a,b],$ *then*

$$\left| f\left(e^{is}\right) \left[u\left(b\right) - u\left(a\right)\right] - \int_a^b f\left(e^{it}\right) du\left(t\right) \right| \qquad (2.34)$$

$$\leq 2L \left[\sin\left(\frac{b-s}{2}\right) u\left(b\right) - \sin\left(\frac{s-a}{2}\right) u\left(a\right)\right]$$

$$+ L \int_a^b \operatorname{sgn}\left(s - t\right) \cos\left(\frac{s-t}{2}\right) u\left(t\right) dt$$

for any $s \in [a,b].$
In particular, we have

$$\left| f\left(e^{\frac{a+b}{2}i}\right) \left[u\left(b\right) - u\left(a\right)\right] - \int_a^b f\left(e^{it}\right) du\left(t\right) \right| \qquad (2.35)$$

$$\leq 2L \sin\left(\frac{b-a}{4}\right) \left[u\left(b\right) - u\left(a\right)\right]$$

$$+ L \int_a^b \operatorname{sgn}\left(\frac{a+b}{2} - t\right) \cos\left(\frac{\frac{a+b}{2} - t}{2}\right) u\left(t\right) dt.$$

Proof. It is well known that if $p : [a,b] \to \mathbb{C}$ is a continuous function and $v : [a,b] \to \mathbb{R}$ is monotonic nondecreasing on $[a,b],$ then the Riemann–Stieltjes integral $\int_a^b p\left(t\right) dv\left(t\right)$ exists and the following inequality holds

$$\left| \int_a^b p\left(t\right) dv\left(t\right) \right| \leq \int_a^b \left|p\left(t\right)\right| dv\left(t\right). \qquad (2.36)$$

Utilizing the property (2.36), we have from (2.7) that

$$
\left| f\left(e^{is}\right)\left[u\left(b\right) - u\left(a\right)\right] - \int_a^b f\left(e^{it}\right) du\left(t\right) \right| \tag{2.37}
$$

$$
= \left| \int_a^b \left[f\left(e^{is}\right) - f\left(e^{it}\right) \right] du\left(t\right) \right|
$$

$$
\leq \int_a^b \left| f\left(e^{is}\right) - f\left(e^{it}\right) \right| du\left(t\right) \leq L \int_a^b \left| e^{is} - e^{it} \right| du\left(t\right)
$$

for any $s \in [a, b]$.

Since, by (2.10), $\left| e^{is} - e^{it} \right| = 2 \left| \sin\left(\frac{s-t}{2}\right) \right|$ for any $t, s \in \mathbb{R}$, then

$$
\int_a^b \left| e^{is} - e^{it} \right| du\left(t\right) \tag{2.38}
$$

$$
= 2 \int_a^b \left| \sin\left(\frac{s-t}{2}\right) \right| du\left(t\right)
$$

$$
= 2 \left[\int_a^s \sin\left(\frac{s-t}{2}\right) du\left(t\right) + \int_s^b \sin\left(\frac{t-s}{2}\right) du\left(t\right) \right]
$$

for any $s \in [a, b] \subseteq [0, 2\pi]$.

Utilizing the integration by parts formula for the Riemann–Stieltjes integral, we have

$$
\int_a^s \sin\left(\frac{s-t}{2}\right) du\left(t\right)
$$

$$
= \sin\left(\frac{s-t}{2}\right) u\left(t\right) \Big|_a^s + \frac{1}{2} \int_a^s \cos\left(\frac{s-t}{2}\right) u\left(t\right) dt
$$

$$
= -\sin\left(\frac{s-a}{2}\right) u\left(a\right) + \frac{1}{2} \int_a^s \cos\left(\frac{s-t}{2}\right) u\left(t\right) dt
$$

and

$$
\int_s^b \sin\left(\frac{t-s}{2}\right) du\left(t\right)
$$

$$
= \sin\left(\frac{t-s}{2}\right) u\left(t\right) \Big|_s^b - \frac{1}{2} \int_s^b \cos\left(\frac{t-s}{2}\right) u\left(t\right) dt
$$

$$
= \sin\left(\frac{b-s}{2}\right) u\left(b\right) - \frac{1}{2} \int_s^b \cos\left(\frac{t-s}{2}\right) u\left(t\right) dt,
$$

which, by (2.38), produce the equality

$$\int_a^b \left| e^{is} - e^{it} \right| du(t) \tag{2.39}$$

$$= 2 \left[\sin\left(\frac{b-s}{2}\right) u(b) - \sin\left(\frac{s-a}{2}\right) u(a) \right]$$

$$+ \int_a^s \cos\left(\frac{s-t}{2}\right) u(t) \, dt - \int_s^b \cos\left(\frac{t-s}{2}\right) u(t) \, dt$$

$$= 2 \left[\sin\left(\frac{b-s}{2}\right) u(b) - \sin\left(\frac{s-a}{2}\right) u(a) \right]$$

$$+ \int_a^b \operatorname{sgn}(s-t) \cos\left(\frac{s-t}{2}\right) u(t) \, dt.$$

Utilizing (2.37) we deduce the desired result (2.34).

Remark 8 *We remark that if $a = 0$ and $b = 2\pi$, then we get from (2.32) and (2.33) that*

$$\left| f\left(e^{is}\right) \left[u(2\pi) - u(0) \right] - \int_0^{2\pi} f\left(e^{it}\right) du(t) \right| \tag{2.40}$$

$$\leq 2L \sin\left(\frac{s}{2}\right) \left[u(2\pi) - u(0) \right]$$

$$+ L \int_0^{2\pi} \operatorname{sgn}(s-t) \cos\left(\frac{s-t}{2}\right) u(t) \, dt$$

for any $s \in [a, b]$.

In particular, we have

$$\left| f(-1) \left[u(2\pi) - u(0) \right] - \int_0^{2\pi} f\left(e^{it}\right) du(t) \right| \tag{2.41}$$

$$\leq \sqrt{2} L \left[u(2\pi) - u(0) \right]$$

$$+ L \int_0^{2\pi} \operatorname{sgn}(\pi - t) \sin\left(\frac{t}{2}\right) u(t) \, dt.$$

Corollary 2.6 (Dragomir 2015, [19]) *Assume that f and u are as in Theorem 2.5, then for any $[a, b] \subset [0, 2\pi]$ with*

$0 < b - a \leq \pi$ *we have the sequence of inequalities*

$$\left| f\left(e^{is}\right) [u(b) - u(a)] - \int_a^b f\left(e^{it}\right) du(t) \right| \tag{2.42}$$

$$\leq 2L \left[\sin\left(\frac{b-s}{2}\right) u(b) - \sin\left(\frac{s-a}{2}\right) u(a) \right]$$

$$+ L \int_a^b \operatorname{sgn}(s-t) \cos\left(\frac{s-t}{2}\right) u(t)\, dt$$

$$\leq 2L \left[\sin\left(\frac{b-s}{2}\right) [u(b) - u(s)] + \sin\left(\frac{s-a}{2}\right) [u(s) - u(a)] \right]$$

$$=: B(s)$$

where

$$B(s) \leq 2L$$

$$\times \begin{cases} \sin\left[\frac{1}{4}(b-a) + \frac{1}{2}\left|s - \frac{a+b}{2}\right|\right] [u(b) - u(a)] \\[2ex] 2\sin\left(\frac{b-a}{4}\right) \cos\left(\frac{s-\frac{a+b}{2}}{2}\right) \left[\frac{u(b)-u(a)}{2} + \left|u(s) - \frac{u(b)+u(a)}{2}\right|\right] \end{cases}$$

for any $s \in [a,b]$.

In particular, we have

$$\left| f\left(e^{\frac{a+b}{2}i}\right) [u(b) - u(a)] - \int_a^b f\left(e^{it}\right) du(t) \right| \tag{2.43}$$

$$\leq 2L \sin\left(\frac{b-a}{4}\right) [u(b) - u(a)]$$

$$+ L \int_a^b \operatorname{sgn}\left(\frac{a+b}{2} - t\right) \cos\left(\frac{\frac{a+b}{2} - t}{2}\right) u(t)\, dt$$

$$=: M.$$

where

$$M \leq 2L \sin\left(\frac{b-a}{4}\right) [u(b) - u(a)].$$

Proof. Since $0 < b - a \leq \pi$, then $\frac{|s-t|}{2} \leq \frac{\pi}{2}$ for $s, t \in [a,b]$.

Utilizing the fact that u is monotonic nondecreasing on $[a, b]$ and $\cos\left(\frac{|s-t|}{2}\right) \geq 0$ for $s, t \in [a, b]$, then

$$\int_a^s \cos\left(\frac{s-t}{2}\right) u(t)\, dt \leq u(s) \int_a^s \cos\left(\frac{s-t}{2}\right) dt \quad (2.44)$$

$$= 2u(s) \sin\left(\frac{s-a}{2}\right)$$

and

$$\int_s^b \cos\left(\frac{s-t}{2}\right) u(t)\, dt \geq u(s) \int_s^b \cos\left(\frac{s-t}{2}\right) dt$$

$$= 2u(s) \sin\left(\frac{b-s}{2}\right)$$

i.e.,

$$-\int_s^b \cos\left(\frac{s-t}{2}\right) u(t)\, dt \leq -2u(s) \sin\left(\frac{b-s}{2}\right). \quad (2.45)$$

Summing (2.44) with (2.45) we deduce that

$$\int_a^b \operatorname{sgn}(s-t) \cos\left(\frac{s-t}{2}\right) u(t)\, dt$$

$$\leq 2u(s) \sin\left(\frac{s-a}{2}\right) - 2u(s) \sin\left(\frac{b-s}{2}\right)$$

giving that

$$2L\left[\sin\left(\frac{b-s}{2}\right) u(b) - \sin\left(\frac{s-a}{2}\right) u(a)\right]$$

$$+ L \int_a^b \operatorname{sgn}(s-t) \cos\left(\frac{s-t}{2}\right) u(t)\, dt$$

$$\leq 2L\left[\sin\left(\frac{b-s}{2}\right)[u(b) - u(s)] + \sin\left(\frac{s-a}{2}\right)[u(s) - u(a)]\right]$$

which proves the second inequality in (2.42).

The bounds for $B(s)$ follow from the elementary property stating that

$$\alpha x + \beta y \leq \max\{\alpha, \beta\}(x + y)$$

where $\alpha, \beta, x, y \geq 0$. The details are omitted.

2.3   A QUADRATURE RULE

We consider the following partition of the interval $[a, b]$

$$\Delta_n : a = x_0 < x_1 < \ldots < x_{n-1} < x_n = b$$

and the intermediate points $\xi_k \in [x_k, x_{k+1}]$ where $0 \leq k \leq n - 1$. Define $h_k := x_{k+1} - x_k$, $0 \leq k \leq n - 1$ and $\nu(\Delta_n) = \max\{h_k : 0 \leq k \leq n - 1\}$ the norm of the partition Δ_n.

For the continuous function $f : \mathcal{C}(0, 1) \to \mathbb{C}$ and the function $u : [a, b] \subseteq [0, 2\pi] \to \mathbb{C}$ of bounded variation on $[a, b]$, define the quadrature rule

$$O_n(f, u, \Delta_n, \xi) := \sum_{k=0}^{n-1} f\left(e^{i\xi_k}\right) [u(x_{k+1}) - u(x_k)] \quad (2.46)$$

and the remainder $R_n(f, u, \Delta_n, \xi)$ in approximating the Riemann–Stieltjes integral $\int_a^b f(e^{it}) \, du(t)$ by $O_n(f, u, \Delta_n, \xi)$. Then we have

$$\int_a^b f\left(e^{it}\right) du(t) = O_n(f, u, \Delta_n, \xi) + R_n(f, u, \Delta_n, \xi). \quad (2.47)$$

The following result provides *a priori* bounds for $R_n(f, u, \Delta_n, \xi)$ in several instances of f and u as above.

Proposition 2.7 (Dragomir 2015, [19]) *Assume that $f : \mathcal{C}(0, 1) \to \mathbb{C}$ satisfies the following Hölder-type condition*

$$|f(z) - f(w)| \leq H |z - w|^r$$

for any $w, z \in \mathcal{C}(0, 1)$, where $H > 0$ and $r \in (0, 1]$ are given. If $[a, b] \subseteq [0, 2\pi]$ and the function $u : [a, b] \to \mathbb{C}$ is of bounded variation on $[a, b]$, then for any partition $\Delta_n : a = x_0 < x_1 < \ldots < x_{n-1} < x_n = b$ with the norm $\nu(\Delta_n) \leq \pi$ we *have the error bound*

$$|R_n(f, u, \Delta_n, \xi)| \qquad (2.48)$$

$$\leq 2^r H \sum_{k=0}^{n-1} \sin^r \left[\frac{1}{4}(x_{k+1} - x_k) + \frac{1}{2} \left| \xi_k - \frac{x_k + x_{k+1}}{2} \right| \right] \bigvee_{x_k}^{x_{k+1}} (u)$$

$$\leq 2^r H \sum_{k=0}^{n-1} \sin^r \left[\frac{1}{2}(x_{k+1} - x_k) \right] \bigvee_{x_k}^{x_{k+1}} (u)$$

$$\leq H \sum_{k=0}^{n-1} (x_{k+1} - x_k)^r \bigvee_{x_k}^{x_{k+1}} (u) \leq H \nu^r (\Delta_n) \bigvee_{a}^{b} (u)$$

for any intermediate points $\xi_k \in [x_k, x_{k+1}]$ where $0 \leq k \leq n-1$.

Proof. Since $\nu(\Delta_n) \leq \pi$, then on writing inequality (2.13) on each interval $[x_k, x_{k+1}]$ and for any intermediate points $\xi_k \in [x_k, x_{k+1}]$ where $0 \leq k \leq n-1$, we have

$$\left| f\left(e^{i\xi_k}\right) [u(x_{k+1}) - u(x_k)] - \int_{x_k}^{x_{k+1}} f\left(e^{it}\right) du(t) \right| \qquad (2.49)$$

$$\leq 2^r H \sin^r \left[\frac{1}{4}(x_{k+1} - x_k) + \frac{1}{2} \left| \xi_k - \frac{x_k + x_{k+1}}{2} \right| \right] \bigvee_{x_k}^{x_{k+1}} (u)$$

$$\leq 2^r H \sin^r \left[\frac{1}{2}(x_{k+1} - x_k) \right] \bigvee_{x_k}^{x_{k+1}} (u) \leq H(x_{k+1} - x_k)^r \bigvee_{x_k}^{x_{k+1}} (u)$$

where for the last inequality we have used the fact that $\sin x \leq x$ for $x \in \left[0, \frac{\pi}{2}\right]$.

Summing over k from 0 to $n-1$ in (2.49) and utilizing the generalized triangle inequality, we deduce the first part of (2.48). The second part is obvious.

Corollary 2.8 (Dragomir 2015, [19]) *Assume that f, u and Δ_n are as in Theorem 2.7. Define the midpoint trapezoid-type quadrature rule by*

$$T_n(f, u, \Delta_n) := \sum_{k=0}^{n-1} f\left(e^{\frac{x_{k+1}+x_k}{2} i}\right) [u(x_{k+1}) - u(x_k)] \qquad (2.50)$$

and the error $E_n(f, u, \Delta_n)$ by

$$\int_a^b f\left(e^{it}\right) du(t) = T_n(f, u, \Delta_n) + E_n(f, u, \Delta_n). \quad (2.51)$$

Then we have the error bounds

$$|E_n(f, u, \Delta_n)| \quad (2.52)$$

$$\leq 2^r H \sum_{k=0}^{n-1} \sin^r\left[\frac{1}{4}(x_{k+1} - x_k)\right] \bigvee_{x_k}^{x_{k+1}}(u)$$

$$\leq \frac{1}{2^r} H \sum_{k=0}^{n-1}(x_{k+1} - x_k)^r \bigvee_{x_k}^{x_{k+1}}(u) \leq \frac{1}{2^r} H \nu^r(\Delta_n) \bigvee_a^b(u).$$

The case of both integrator and integrand being Lipschitzian is incorporated in the following result:

Proposition 2.9 (Dragomir 2015, [19]) *Assume that f : $\mathcal{C}(0,1) \to \mathbb{C}$ is Lipschitzian with the constant $L > 0$ on the circle $\mathcal{C}(0,1)$. If $[a, b] \subseteq [0, 2\pi]$ and the function $u : [a, b] \to \mathbb{C}$ is Lipschitzian with the constant $K > 0$ on $[a, b]$, then for any partition $\Delta_n : a = x_0 < x_1 < ... < x_{n-1} < x_n = b$ we have the error bound*

$$|R_n(f, u, \Delta_n, \xi)| \quad (2.53)$$

$$\leq 4LK \sum_{k=0}^{n-1}\left[\sin^2\left(\frac{\xi_k - x_k}{4}\right) + \sin^2\left(\frac{x_{k+1} - \xi_k}{4}\right)\right]$$

$$\leq 8LK \sum_{k=0}^{n-1} \sin^2\left(\frac{x_{k+1} - x_k}{4}\right) \leq \frac{1}{2}LK \sum_{k=0}^{n-1}(x_{k+1} - x_k)^2$$

$$\leq \frac{1}{2}LK(b-a)\nu(\Delta_n)$$

for any intermediate points $\xi_k \in [x_k, x_{k+1}]$ where $0 \leq k \leq n-1$.

In particular, we have

$$|E_n(f, u, \Delta_n)| \leq 8LK \sum_{k=0}^{n-1} \sin^2\left(\frac{x_{k+1} - x_k}{8}\right) \quad (2.54)$$

$$\leq \frac{1}{8}LK \sum_{k=0}^{n-1}(x_{k+1} - x_k)^2 \leq \frac{1}{8}LK(b-a)\nu(\Delta_n).$$

The proof follows by Theorem 2.3 and the details are omitted.

Proposition 2.10 (Dragomir 2015, [19]) *Assume that* $f :$ $\mathcal{C}(0,1) \to \mathbb{C}$ *is Lipschitzian with the constant* $L > 0$ *on the circle* $\mathcal{C}(0,1)$. *If* $[a,b] \subseteq [0,2\pi]$ *and the function* $u : [a,b] \to \mathbb{R}$ *is monotonic nondecreasing on* $[a,b]$, *then for any partition* $\Delta_n : a = x_0 < x_1 < ... < x_{n-1} < x_n = b$ *with the norm* $\nu(\Delta_n) \leq \pi$ *we have the error bound*

$$|R_n(f, u, \Delta_n, \xi)| \tag{2.55}$$

$$\leq 2L \sum_{k=0}^{n-1} \left[\sin\left(\frac{x_{k+1} - \xi_k}{2}\right) u(x_{k+1}) - \sin\left(\frac{\xi_k - x_k}{2}\right) u(x_k) \right]$$

$$+ L \sum_{k=0}^{n-1} \int_{x_k}^{x_{k+1}} \operatorname{sgn}(\xi_k - t) \cos\left(\frac{\xi_k - t}{2}\right) u(t)\, dt$$

$$\leq 2L \sum_{k=0}^{n-1} \left[\sin\left(\frac{x_{k+1} - \xi_k}{2}\right) [u(x_{k+1}) - u(\xi_k)] \right.$$

$$\left. + \sin\left(\frac{\xi_k - x_k}{2}\right) [u(\xi_k) - u(x_k)] \right]$$

$$\leq 2L \sum_{k=0}^{n-1} \sin\left[\frac{1}{4}(x_{k+1} - x_k) + \frac{1}{2}\left|\xi_k - \frac{x_k + x_{k+1}}{2}\right|\right]$$

$$[u(x_{k+1}) - u(x_k)]$$

$$\leq 2L \sum_{k=0}^{n-1} \sin\left[\frac{1}{2}(x_{k+1} - x_k)\right] [u(x_{k+1}) - u(x_k)]$$

$$\leq L \sum_{k=0}^{n-1} (x_{k+1} - x_k) [u(x_{k+1}) - u(x_k)] \leq \nu(\Delta_n)$$

$$L[u(b) - u(a)]$$

for any intermediate points $\xi_k \in [x_k, x_{k+1}]$ *where* $0 \leq k \leq n - 1$.

In particular, we have

$$|E_n (f, u, \Delta_n)| \tag{2.56}$$

$$\leq 2L \sum_{k=0}^{n-1} \sin\left(\frac{x_{k+1} - x_k}{4}\right) [u(x_{k+1}) - u(x_k)]$$

$$+ L \sum_{k=0}^{n-1} \int_{x_k}^{x_{k+1}} \text{sgn}\left(\frac{x_k + x_{k+1}}{2} - t\right) \cos\left(\frac{\frac{x_k + x_{k+1}}{2} - t}{2}\right) u(t)\, dt$$

$$\leq 2L \sum_{k=0}^{n-1} \sin\left(\frac{x_{k+1} - x_k}{4}\right) [u(x_{k+1}) - u(x_k)]$$

$$\leq \frac{1}{2} L \sum_{k=0}^{n-1} (x_{k+1} - x_k) [u(x_{k+1}) - u(x_k)]$$

$$\leq \frac{1}{2} L \nu(\Delta_n) [u(b) - u(a)].$$

The proof follows by Corollary 2.6 and the details are omitted.

2.4 APPLICATIONS FOR FUNCTIONS OF UNITARY OPERATORS

We consider the following partition of the interval $[a, b]$

$$\Delta_n : 0 = \lambda_0 < \lambda_1 < \ldots < \lambda_{n-1} < \lambda_n = 2\pi$$

and the intermediate points $\xi_k \in [\lambda_k, \lambda_{k+1}]$ where $0 \leq k \leq n - 1$. Define $h_k := \lambda_{k+1} - \lambda_k$, $0 \leq k \leq n - 1$ and $\nu(\Delta_n) = \max\{h_k : 0 \leq k \leq n - 1\}$, the norm of the partition Δ_n.

If U is a unitary operator on the Hilbert space H and $\{E_\lambda\}_{\lambda \in [0, 2\pi]}$, the spectral family of U, then we can introduce the following sums

$$O_n (f, U, \Delta_n, \xi; x, y) := \sum_{k=0}^{n-1} f\left(e^{i\xi_k}\right) \langle (E_{\lambda_{k+1}} - E_{\lambda_k}) x, y \rangle$$

$$\tag{2.57}$$

and

$$T_n\left(f, U, \Delta_n; x, y\right) := \sum_{k=0}^{n-1} f\left(e^{\frac{\lambda_{k+1}+\lambda_k}{2}i}\right) \left\langle \left(E_{\lambda_{k+1}} - E_{\lambda_k}\right) x, y \right\rangle$$

$$(2.58)$$

where $x, y \in H$.

Theorem 2.11 (Dragomir 2015, [19]) *With the above assumptions for* U, $\{E_\lambda\}_{\lambda \in [0,2\pi]}$, Δ_n *with* $\nu\left(\Delta_n\right) \leq \pi$ *and if* $f : \mathcal{C}\left(0,1\right) \to \mathbb{C}$ *satisfies the Hölder-type condition* $\left|f\left(z\right) - f\left(w\right)\right| \leq H\left|z - w\right|^r$ *for any* $w, z \in \mathcal{C}\left(0,1\right)$, *where* $H > 0$ *and* $r \in (0,1]$ *are given, then we have the representation*

$$\left\langle f\left(U\right) x, y \right\rangle = O_n\left(f, U, \Delta_n, \xi; x, y\right) + R_n\left(f, U, \Delta_n, \xi; x, y\right)$$

$$(2.59)$$

with the error $R_n\left(f, U, \Delta_n, \xi; x, y\right)$ *satisfying the bounds*

$$\left|R_n\left(f, U, \Delta_n, \xi; x, y\right)\right| \qquad (2.60)$$

$$\leq 2^r H \sum_{k=0}^{n-1} \sin^r \left[\frac{1}{4}\left(\lambda_{k+1} - \lambda_k\right) + \frac{1}{2}\left|\xi_k - \frac{\lambda_k + \lambda_{k+1}}{2}\right|\right] \bigvee_{\lambda_k}^{\lambda_{k+1}} \left(\left\langle E_{(\cdot)} x, y \right\rangle\right)$$

$$\leq 2^r H \sum_{k=0}^{n-1} \sin^r \left[\frac{1}{2}\left(\lambda_{k+1} - \lambda_k\right)\right] \bigvee_{\lambda_k}^{\lambda_{k+1}} \left(\left\langle E_{(\cdot)} x, y \right\rangle\right)$$

$$\leq H \sum_{k=0}^{n-1} \left(\lambda_{k+1} - \lambda_k\right)^r \bigvee_{\lambda_k}^{\lambda_{k+1}} \left(\left\langle E_{(\cdot)} x, y \right\rangle\right) \leq H\nu^r\left(\Delta_n\right) \bigvee_0^{2\pi} \left(\left\langle E_{(\cdot)} x, y \right\rangle\right)$$

$$\leq H\nu^r\left(\Delta_n\right) \|x\| \|y\|$$

for any $x, y \in H$ *and the intermediate points* $\xi_k \in [\lambda_k, \lambda_{k+1}]$ *where* $0 \leq k \leq n - 1$.

In particular we have

$$\left\langle f\left(U\right) x, y \right\rangle = T_n\left(f, U, \Delta_n; x, y\right) + E_n\left(f, U, \Delta_n; x, y\right) \quad (2.61)$$

with the error

$$|E_n(f, U, \Delta_n; x, y)| \tag{2.62}$$

$$\leq 2^r H \sum_{k=0}^{n-1} \sin^r \left[\frac{1}{4} (\lambda_{k+1} - \lambda_k) \right] \bigvee_{\lambda_k}^{\lambda_{k+1}} \left(\left\langle E_{(\cdot)} x, y \right\rangle \right)$$

$$\leq \frac{1}{2^r} H \sum_{k=0}^{n-1} (\lambda_{k+1} - \lambda_k)^r \bigvee_{\lambda_k}^{\lambda_{k+1}} \left(\left\langle E_{(\cdot)} x, y \right\rangle \right)$$

$$\leq \frac{1}{2^r} H \nu^r (\Delta_n) \bigvee_{0}^{2\pi} \left(\left\langle E_{(\cdot)} x, y \right\rangle \right) \leq \frac{1}{2^r} H \nu^r (\Delta_n) \|x\| \|y\|$$

for any $x, y \in H$.

Proof. For given $x, y \in H$, define the function $u(\lambda) := \langle E_\lambda x, y \rangle$, $\lambda \in [0, 2\pi]$. We know that, see Theorem 1.1, u is of bounded variation and

$$\bigvee_{0}^{2\pi} (u) =: \bigvee_{0}^{2\pi} \left(\left\langle E_{(\cdot)} x, y \right\rangle \right) \leq \|x\| \|y\|. \tag{2.63}$$

Now, applying Proposition 2.7 to the spectral representation of unitary operators we deduce the desired result (2.59) with the error bound (2.60). The details are omitted.

Remark 9 *In the case when the partition reduces to the whole interval $[0, 2\pi]$, then utilizing the inequality (2.11) we deduce the bound*

$$\left| f \left(e^{is} \right) \langle x, y \rangle - \langle f(U) x, y \rangle \right| \leq 2^r H \bigvee_{0}^{2\pi} \left(\left\langle E_{(\cdot)} x, y \right\rangle \right)$$

$$\leq 2^r H \|x\| \|y\| \tag{2.64}$$

for any $s \in [0, 2\pi]$ and any vectors $x, y \in H$.
In the case when the division is

$$\Delta_2 : 0 = \lambda_0 < \lambda_1 = \pi < \lambda_2 = 2\pi$$

and we take the intermediate points $u \in [0, \pi]$ *and* $v \in [\pi, 2\pi]$, *then we get from Theorem 2.11 that*

$$
\left| f\left(e^{iu}\right) \langle E_\pi x, y \rangle + f\left(e^{iv}\right) \langle (1_H - E_\pi) x, y \rangle - \langle f(U) x, y \rangle \right|
$$

$$
\leq 2^r H \left[\sin^r \left[\frac{1}{4}\pi + \frac{1}{2} \left| u - \frac{\pi}{2} \right| \right] \bigvee_0^\pi \left(\langle E_{(\cdot)} x, y \rangle \right) \right.
$$

$$
\left. + \sin^r \left[\frac{1}{4}\pi + \frac{1}{2} \left| v - \frac{3\pi}{2} \right| \right] \bigvee_\pi^{2\pi} \left(\langle E_{(\cdot)} x, y \rangle \right) \right] \quad (2.65)
$$

for any vectors $x, y \in H$.

The best inequality we can get from (2.66) is obtained for $u = \frac{\pi}{2}$ *and* $v = \frac{3\pi}{2}$, *namely*

$$
\left| f(i) \langle E_\pi x, y \rangle + f(-i) \langle (1_H - E_\pi) x, y \rangle - \langle f(U) x, y \rangle \right|
$$

$$
(2.66)
$$

$$
\leq 2^{\frac{r}{2}} H \bigvee_0^{2\pi} \left(\langle E_{(\cdot)} x, y \rangle \right) \leq 2^{\frac{r}{2}} H \|x\| \|y\|
$$

for any vectors $x, y \in H$.

If U is a unitary operator on the Hilbert space H and $\{E_\lambda\}_{\lambda \in [0, 2\pi]}$, the spectral family of U, then we can introduce the following sums depending only on one vector $x \in H$

$$
\tilde{O}_n (f, U, \Delta_n, \xi; x) := \sum_{k=0}^{n-1} f\left(e^{i\xi_k}\right) \langle (E_{\lambda_{k+1}} - E_{\lambda_k}) x, x \rangle
$$

$$
(2.67)
$$

and

$$
\tilde{T}_n (f, U, \Delta_n; x, y) := \sum_{k=0}^{n-1} f\left(e^{\frac{\lambda_{k+1}+\lambda_k}{2}i}\right) \langle (E_{\lambda_{k+1}} - E_{\lambda_k}) x, x \rangle .
$$

$$
(2.68)
$$

Theorem 2.12 (Dragomir 2015, [19]) *With the above assumptions for* U, $\{E_\lambda\}_{\lambda \in [0, 2\pi]}$, Δ_n *with* $\nu(\Delta_n) \leq \pi$ *and, if*

$f : \mathcal{C}(0,1) \to \mathbb{C}$ *is Lipschitzian with the constant* $L > 0$ *on the circle* $\mathcal{C}(0,1)$*, then we have the representation*

$$\langle f(U) x, x \rangle = \tilde{O}_n (f, U, \Delta_n, \xi; x) + \tilde{R}_n (f, U, \Delta_n, \xi; x) \quad (2.69)$$

with the error $\tilde{R}_n (f, U, \Delta_n, \xi; x)$ *satisfying the bounds*

$$\left| \tilde{R}_n (f, U, \Delta_n, \xi; x) \right| \quad (2.70)$$

$$\leq 2L \sum_{k=0}^{n-1} \left[\sin \left(\frac{\lambda_{k+1} - \xi_k}{2} \right) \langle E_{\lambda_{k+1}} x, x \rangle - \sin \left(\frac{\xi_k - \lambda_k}{2} \right) \right.$$

$$\left. \times \langle E_{\lambda_k} x, x \rangle \right] + L \sum_{k=0}^{n-1} \int_{\lambda_k}^{\lambda_{k+1}} \mathrm{sgn} \, (\xi_k - t) \cos \left(\frac{\xi_k - t}{2} \right) \langle E_t x, x \rangle \, dt$$

$$\leq 2L \sum_{k=0}^{n-1} \left[\sin \left(\frac{\lambda_{k+1} - \xi_k}{2} \right) \left[\langle (E_{\lambda_{k+1}} - E_{\xi_k}) x, x \rangle \right] \right.$$

$$+ \sin \left(\frac{\xi_k - \lambda_k}{2} \right) \langle (E_{\xi_k} - E_{\lambda_k}) x, x \rangle \right]$$

$$\leq 2L \sum_{k=0}^{n-1} \sin \left[\frac{1}{4} (\lambda_{k+1} - \lambda_k) + \frac{1}{2} \left| \xi_k - \frac{\lambda_k + \lambda_{k+1}}{2} \right| \right]$$

$$\langle (E_{\lambda_{k+1}} - E_{\lambda_k}) x, x \rangle$$

$$\leq 2L \sum_{k=0}^{n-1} \sin \left[\frac{1}{2} (\lambda_{k+1} - \lambda_k) \right] \langle (E_{\lambda_{k+1}} - E_{\lambda_k}) x, x \rangle$$

$$\leq L \sum_{k=0}^{n-1} (\lambda_{k+1} - \lambda_k) \langle (E_{\lambda_{k+1}} - E_{\lambda_k}) x, x \rangle \leq \nu (\Delta_n) L \|x\|^2$$

for any $x \in H$ *and the intermediate points* $\xi_k \in [\lambda_k, \lambda_{k+1}]$ *where* $0 \leq k \leq n - 1$.

In particular we have

$$\langle f(U) x, x \rangle = \tilde{T}_n (f, U, \Delta_n; x) + \tilde{E}_n (f, U, \Delta_n; x) \quad (2.71)$$

with the error

$$\left| \tilde{E}_n \left(f, U, \Delta_n; x \right) \right| \tag{2.72}$$

$$\leq 2L \sum_{k=0}^{n-1} \sin \left(\frac{\lambda_{k+1} - \lambda_k}{4} \right) \left\langle \left(E_{\lambda_{k+1}} - E_{\lambda_k} \right) x, x \right\rangle$$

$$+ L \sum_{k=0}^{n-1} \int_{x_k}^{x_{k+1}} \operatorname{sgn} \left(\frac{\lambda_k + \lambda_{k+1}}{2} - t \right) \cos \left(\frac{\frac{\lambda_k + \lambda_{k+1}}{2} - t}{2} \right)$$

$$\times \left\langle E_t x, x \right\rangle dt$$

$$\leq 2L \sum_{k=0}^{n-1} \sin \left(\frac{\lambda_{k+1} - \lambda_k}{4} \right) \left\langle \left(E_{\lambda_{k+1}} - E_{\lambda_k} \right) x, x \right\rangle$$

$$\leq \frac{1}{2} L \sum_{k=0}^{n-1} \left(\lambda_{k+1} - \lambda_k \right) \left\langle \left(E_{\lambda_{k+1}} - E_{\lambda_k} \right) x, x \right\rangle$$

$$\leq \frac{1}{2} L \nu \left(\Delta_n \right) \| x \|^2$$

for any $x \in H$.

The proof follows by Proposition 2.10 applied for the monotonic nondecreasing function $u(t) := \langle E_t x, x \rangle, t \in [0, 2\pi]$.

Remark 10 *We remark that if the partition reduces to the whole interval $[0, 2\pi]$ then we get from (2.40) that*

$$\left| f \left(e^{is} \right) \| x \|^2 - \left\langle f \left(U \right) x, x \right\rangle \right| \tag{2.73}$$

$$\leq 2L \sin \left(\frac{s}{2} \right) \| x \|^2 + L \int_0^{2\pi} \operatorname{sgn} \left(s - t \right) \cos \left(\frac{s - t}{2} \right) \left\langle E_t x, x \right\rangle dt$$

for any $s \in [a, b]$ and $x \in H$.
In particular, we have

$$\left| f \left(-1 \right) \| x \|^2 - \left\langle f \left(U \right) x, x \right\rangle \right| \tag{2.74}$$

$$\leq \sqrt{2} L \| x \|^2 + L \int_0^{2\pi} \operatorname{sgn} \left(\pi - t \right) \sin \left(\frac{t}{2} \right) \left\langle E_t x, x \right\rangle dt$$

for any $x \in H$.

Example 2.1 *In order to provide some simple examples for the inequalities above, we choose two complex functions as follows.*

a) *Consider the power function* $f : \mathbb{C} \setminus \{0\} \to \mathbb{C}$, $f(z) = z^m$ *where m is a nonzero integer. Then, obviously, for any z, w belonging to the unit circle $\mathcal{C}(0, 1)$ we have the inequality*

$$|f(z) - f(w)| \leq |m| \, |z - w|$$

which shows that f is Lipschitzian with the constant $L = |m|$ on the circle $\mathcal{C}(0, 1)$. Then from (2.64), we get for any unitary operator U that

$$\left| e^{ims} \langle x, y \rangle - \langle U^m x, y \rangle \right| \leq 2 |m| \bigvee_0^{2\pi} \left(\langle E_{(\cdot)} x, y \rangle \right)$$

$$\leq 2 |m| \, \|x\| \, \|y\| \qquad (2.75)$$

for any $s \in [0, 2\pi]$ and $x, y \in H$.

Also, from (2.65) and the intermediate points $u \in [0, \pi]$ and $v \in [\pi, 2\pi]$, we have for any unitary operator U

$$\left| e^{imu} \langle E_\pi x, y \rangle + e^{imv} \langle (1_H - E_\pi) x, y \rangle - \langle U^m x, y \rangle \right|$$

$$\tag{2.76}$$

$$\leq 2 |m| \left[\sin \left[\frac{1}{4} \pi + \frac{1}{2} \left| u - \frac{\pi}{2} \right| \right] \bigvee_0^\pi \left(\langle E_{(\cdot)} x, y \rangle \right) \right.$$

$$\left. + \sin \left[\frac{1}{4} \pi + \frac{1}{2} \left| v - \frac{3\pi}{2} \right| \right] \bigvee_\pi^{2\pi} \left(\langle E_{(\cdot)} x, y \rangle \right) \right]$$

for any vectors $x, y \in H$, where $\{E_\lambda\}_{\lambda \in [0, 2\pi]}$ is the spectral family of U.

The best inequality we can get from (2.76) is obtained

for $u = \frac{\pi}{2}$ *and* $v = \frac{3\pi}{2}$, *namely*

$$\left| i^m \left\langle E_\pi x, y \right\rangle + (-i)^m \left\langle (1_H - E_\pi) x, y \right\rangle - \left\langle U^m x, y \right\rangle \right|$$

(2.77)

$$\leq \sqrt{2} \, |m| \bigvee_0^{2\pi} \left(\left\langle E_{(\cdot)} x, y \right\rangle \right) \leq \sqrt{2} \, |m| \, \|x\| \, \|y\| \, ,$$

for any vectors $x, y \in H$.

b) *For* $a \neq \pm 1, 0$ *consider the function* $f : \mathcal{C}(0,1) \to \mathbb{C}$, $f_a(z) = \frac{1}{1-az}$. *Observe that*

$$|f_a(z) - f_a(w)| = \frac{|a| \, |z - w|}{|1 - az| \, |1 - aw|}$$

(2.78)

for any $z, w \in \mathcal{C}(0,1)$.

If $z = e^{it}$ *with* $t \in [0, 2\pi]$, *then we have*

$$\begin{aligned}
|1 - az|^2 &= 1 - 2a \operatorname{Re}(\bar{z}) + a^2 |z|^2 = 1 - 2a \cos t + a^2 \\
&\geq 1 - 2|a| + a^2 = (1 - |a|)^2
\end{aligned}$$

therefore

$$\frac{1}{|1 - az|} \leq \frac{1}{|1 - |a||} \quad and \quad \frac{1}{|1 - aw|} \leq \frac{1}{|1 - |a||}$$

(2.79)

for any $z, w \in \mathcal{C}(0,1)$.

Utilizing (2.78) and (2.79) we deduce

$$|f_a(z) - f_a(w)| \leq \frac{|a|}{(1 - |a|)^2} \, |z - w|$$

(2.80)

for any $z, w \in \mathcal{C}(0,1)$, *showing that the function* f_a *is Lipschitzian with the constant* $L_a = \frac{|a|}{(1-|a|)^2}$ *on the circle* $\mathcal{C}(0,1)$.

Applying the inequality (2.64), we get for any unitary

operator U *that*

$$\left|\left(1 - ae^{is}\right)^{-1} \langle x, y \rangle - \left\langle (1_H - aU)^{-1} x, y \right\rangle\right| \quad (2.81)$$

$$\leq \frac{2|a|}{(1 - |a|)^2} \bigvee_0^{2\pi} \left(\left\langle E_{(\cdot)}x, y \right\rangle\right) \leq \frac{2|a|}{(1 - |a|)^2} \|x\| \|y\|$$

for any $s \in [0, 2\pi]$ *and* $x, y \in H$.

Also, from (2.65) and the intermediate points $u \in [0, \pi]$ *and* $v \in [\pi, 2\pi]$, *we have for any unitary operator* U

$$\left|\left(1 - ae^{iu}\right)^{-1} \langle E_\pi x, y \rangle + \left(1 - ae^{iv}\right)^{-1} \langle (1_H - E_\pi) x, y \rangle\right.$$

$$(2.82)$$

$$\left. - \left\langle (1_H - aU)^{-1} x, y \right\rangle\right|$$

$$\leq \frac{2|a|}{(1 - |a|)^2} \left[\sin\left[\frac{1}{4}\pi + \frac{1}{2}\left|u - \frac{\pi}{2}\right|\right] \bigvee_0^\pi \left(\left\langle E_{(\cdot)}x, y \right\rangle\right)\right.$$

$$\left. + \sin\left[\frac{1}{4}\pi + \frac{1}{2}\left|v - \frac{3\pi}{2}\right|\right] \bigvee_\pi^{2\pi} \left(\left\langle E_{(\cdot)}x, y \right\rangle\right)\right]$$

for any vectors $x, y \in H$, *where* $\{E_\lambda\}_{\lambda \in [0, 2\pi]}$ *is the spectral family of* U.

The best inequality we can get from (2.82) is obtained for $u = \frac{\pi}{2}$ *and* $v = \frac{3\pi}{2}$, *namely*

$$\left|(1 - ai)^{-1} \langle E_\pi x, y \rangle + (1 + ai)^{-1} \langle (1_H - E_\pi) x, y \rangle\right.$$

$$\left. - \left\langle (1_H - aU)^{-1} x, y \right\rangle\right|$$

$$\leq \frac{\sqrt{2}|a|}{(1 - |a|)^2} \bigvee_0^{2\pi} \left(\left\langle E_{(\cdot)}x, y \right\rangle\right) \leq \frac{\sqrt{2}|a|}{(1 - |a|)^2} \|x\| \|y\|$$

$$(2.83)$$

for any vectors $x, y \in H$.

The interested reader may apply the above results for other divisions of the interval $[0, 2\pi]$, for instance

$$\Delta_4 : 0 = \lambda_0 < \lambda_1 = \frac{\pi}{2} < \lambda_2 = \pi < \lambda_3 = \frac{3\pi}{2} < \lambda_4 = 2\pi.$$

However, the details are omitted.

CHAPTER 3

Trapezoid-Type Inequalities

IN THIS CHAPTER we present some results in approximating the Riemann–Stieltjes integral by the trapezoidal rule for continuous complex-valued integrands and various classes of bounded variation integrators. Several applications for functions of unitary operators in Hilbert spaces are provided as well.

3.1 INTRODUCTION

A simple way to approximate the Riemann–Stieltjes integral $\int_a^b f(t) \, du(t)$ is by using the *trapezoidal rule*

$$\frac{f(a) + f(b)}{2} \cdot [u(b) - u(a)] \qquad (3.1)$$

under different assumptions for the *integrand* f and the *integrator* u for which the above integral exists.

A priori error bounds, namely, upper bounds for the quantity

$$\left| \int_a^b f(t) \, du(t) - \frac{f(a) + f(b)}{2} \cdot [u(b) - u(a)] \right|$$

are known for various pairs (f, u) for which the integral $\int_a^b f(t) \, du(t)$ exists. We present here some simple ones.

Theorem 3.1 (Dragomir, 2001, [11]) *Let* $f : [a, b] \to \mathbb{C}$ *be a* $p - H$*-Hölder-type function, that is, it satisfies the condition*

$$|f(x) - f(y)| \le H |x - y|^p \text{ for all } x, y \in [a, b], \quad (3.2)$$

where $H > 0$ *and* $p \in (0, 1]$ *are given, and* $u : [a, b] \to \mathbb{C}$ *is a function of bounded variation on* $[a, b]$. *Then we have the inequality:*

$$\left| \frac{f(a) + f(b)}{2} \cdot [u(b) - u(a)] - \int_a^b f(t) \, du(t) \right| \quad (3.3)$$

$$\le \frac{1}{2^p} H (b - a)^p \bigvee_a^b (u).$$

The constant $C = 1$ *on the right-hand side of (3.3) cannot be replaced by a smaller quantity.*

In the case when u is monotonic nondecreasing, we have the following result as well:

Theorem 3.2 (Dragomir, 2011, [12]) *Let* $f : [a, b] \to \mathbb{C}$ *be a* $p - H$*-Hölder-type mapping where* $H > 0$ *and* $p \in (0, 1]$ *are given, and* $u : [a, b] \to \mathbb{R}$ *a monotonic nondecreasing function on* $[a, b]$. *Then we have the inequality:*

$$\left| \frac{f(a) + f(b)}{2} \cdot [u(b) - u(a)] - \int_a^b f(t) \, du(t) \right| \quad (3.4)$$

$$\le \frac{1}{2} H \left\{ (b - a)^p [u(b) - u(a)] \right.$$

$$\left. - p \int_a^b \left[\frac{(b - t)^{1-p} - (t - a)^{1-p}}{(b - t)^{1-p} (t - a)^{1-p}} \right] u(t) \, dt \right\}$$

$$\le \frac{1}{2^p} H (b - a)^p [u(b) - u(a)].$$

The inequalities in (3.4) are sharp.

The case when both the integrand and the integrator are of bounded variation is as follows:

Theorem 3.3 (Dragomir, 2011, [12]) *Let f, $u : [a, b] \to \mathbb{C}$ be of bounded variation on $[a, b]$. If one of them is continuous on $[a, b]$, then the Riemann–Stieltjes integral $\int_a^b f(t)\, du(t)$ exists and we have the inequality*

$$\left| \frac{f(a) + f(b)}{2} [u(b) - u(a)] - \int_a^b f(t)\, du(t) \right| \leq \frac{1}{2} \bigvee_a^b (f) \bigvee_a^b (u).$$
$$(3.5)$$

The constant $\frac{1}{2}$ is the best possible in (3.5).

For other results of this type, see [12], where applications for functions of self-adjoint operators on complex Hilbert spaces are given as well.

Motivated by the above facts, we consider in this chapter the problem of approximating the Riemann–Stieltjes integral $\int_a^b f(e^{is})\, du(s)$ by the trapezoidal rule

$$\frac{f(e^{ib}) + f(e^{ia})}{2} [u(b) - u(a)]$$

for continuous the complex-valued function $f : \mathcal{C}(0, 1) \to \mathbb{C}$ defined on the complex unit circle $\mathcal{C}(0, 1)$ and various subclasses of functions $u : [a, b] \subseteq [0, 2\pi] \to \mathbb{C}$ of bounded variation.

3.2 TRAPEZOID-TYPE INEQUALITIES

We have the following result.

Theorem 3.4 (Dragomir 2016, [21]) *Assume that $f : \mathcal{C}(0, 1) \to \mathbb{C}$ satisfies the following Hölder-type condition*

$$|f(z) - f(w)| \leq H |z - w|^r \qquad (3.6)$$

for any $w, z \in \mathcal{C}(0, 1)$, where $H > 0$ and $r \in (0, 1]$ are given.

If $[a,b] \subseteq [0,2\pi]$ and the function $u : [a,b] \to \mathbb{C}$ is of bounded variation on $[a,b]$, then

$$\left| \frac{f\left(e^{ib}\right) + f\left(e^{ia}\right)}{2} [u(b) - u(a)] - \int_a^b f\left(e^{is}\right) du(s) \right| \quad (3.7)$$

$$\leq 2^{r-1} H \max_{s \in [a,b]} B_r(a,b;s) \bigvee_a^b (u) \leq \frac{1}{2^r} H (b-a)^r \bigvee_a^b (u)$$

for any $t \in [a,b]$, where the bound $B_r(a,b;s)$ is given by

$$B_r(a,b;s) := \sin^r \left(\frac{b-s}{2} \right) + \sin^r \left(\frac{s-a}{2} \right) \quad (3.8)$$

$$\leq \frac{1}{2^r} [(b-s)^r + (s-a)^r].$$

Moreover, if $f : \mathcal{C}(0,1) \to \mathbb{C}$ is Lipschitzian with the constant $K > 0$, then

$$\left| \frac{f\left(e^{ib}\right) + f\left(e^{ia}\right)}{2} [u(b) - u(a)] - \int_a^b f\left(e^{is}\right) du(s) \right| \quad (3.9)$$

$$\leq 2K \sin \left(\frac{b-a}{4} \right) \bigvee_a^b (u) \leq \frac{1}{2} K (b-a) \bigvee_a^b (u).$$

The constant 2 in the first inequality in (3.9) is best possible in the sense that it cannot be replaced by a smaller quantity.

Proof. We have the equality

$$\frac{f\left(e^{ib}\right) + f\left(e^{ia}\right)}{2} [u(b) - u(a)] - \int_a^b f\left(e^{is}\right) du(s) \quad (3.10)$$

$$= \int_a^b \left[\frac{f\left(e^{ib}\right) + f\left(e^{ia}\right)}{2} - f\left(e^{is}\right) \right] du(s).$$

It is known that if $p : [c,d] \to \mathbb{C}$ is a continuous function and $v : [c,d] \to \mathbb{C}$ is of bounded variation, then the Riemann–Stieltjes integral $\int_c^d p(t) dv(t)$ exists and the following inequality holds

$$\left| \int_c^d p(t) dv(t) \right| \leq \max_{t \in [c,d]} |p(t)| \bigvee_c^d (v). \quad (3.11)$$

Taking the modulus in the equality (3.10) and utilizing the property (3.11) we deduce

$$\left| \frac{f\left(e^{ib}\right) + f\left(e^{ia}\right)}{2} \left[u\left(b\right) - u\left(a\right)\right] - \int_a^b f\left(e^{is}\right) du\left(s\right) \right| \qquad (3.12)$$

$$\leq \left| \int_a^b \left[\frac{f\left(e^{ib}\right) + f\left(e^{ia}\right)}{2} - f\left(e^{is}\right) \right] du\left(s\right) \right| \qquad (3.13)$$

$$\leq \max_{s\in[a,b]} \left| \frac{f\left(e^{ib}\right) + f\left(e^{ia}\right)}{2} - f\left(e^{is}\right) \right| \bigvee_a^b\left(u\right)$$

$$\leq \frac{1}{2} \max_{s\in[a,b]} \left[\left| f\left(e^{ib}\right) - f\left(e^{is}\right) \right| + \left| f\left(e^{is}\right) - f\left(e^{ia}\right) \right| \right] \bigvee_a^b\left(u\right)$$

$$\leq \frac{1}{2} H \max_{s\in[a,b]} \left[\left| e^{ib} - e^{is} \right|^r + \left| e^{is} - e^{ia} \right|^r \right] \bigvee_a^b\left(u\right).$$

Since

$$\left| e^{is} - e^{it} \right|^2 = \left| e^{is} \right|^2 - 2\operatorname{Re}\left(e^{i(s-t)}\right) + \left| e^{it} \right|^2$$

$$= 2 - 2\cos\left(s - t\right) = 4\sin^2\left(\frac{s-t}{2}\right)$$

for any $t, s \in \mathbb{R}$, then

$$\left| e^{is} - e^{it} \right|^r = 2^r \left| \sin\left(\frac{s-t}{2}\right) \right|^r \qquad (3.14)$$

for any $t, s \in \mathbb{R}$.

For $[a, b] \subseteq [0, 2\pi]$ we have

$$\left| e^{ib} - e^{is} \right|^r = 2^r \sin^r\left(\frac{b-s}{2}\right)$$

and

$$\left| e^{is} - e^{ia} \right|^r = 2^r \sin^r\left(\frac{s-a}{2}\right)$$

for any $s \in [a, b]$.

Utilizing the inequality (3.12) we deduce the first inequality in (3.7).

By the elementary inequality $\sin x \leq x$ for $x \in [0, \pi]$ we have the inequality (3.8).

Consider the function $\varphi : [a, b] \rightarrow \mathbb{R}$, $\varphi(s) = (b - s)^r + (s - a)^r$. We have

$$\varphi'(s) = r(s - a)^{r-1} - r(b - s)^{r-1} = r\frac{(b - s)^{1-r} - (s - a)^{1-r}}{(b - s)^{1-r}(s - a)^{1-r}}$$

and

$$\varphi''(s) = r(r - 1)\left[(s - a)^{r-2} + (b - s)^{r-2}\right]$$

for any $s \in (a, b)$. We observe that $\varphi'(s) = 0$ iff $s = \frac{a+b}{2}$, $\varphi'(s) > 0$ for $s \in \left(a, \frac{a+b}{2}\right)$ and $\varphi'(s) < 0$ for $s \in \left(\frac{a+b}{2}, b\right)$ which shows that the function φ is strictly increasing on $\left(a, \frac{a+b}{2}\right)$ and strictly decreasing on $\left(\frac{a+b}{2}, b\right)$. Since $\varphi''(s) < 0$ for any $s \in (a, b)$, the function φ is strictly concave on $[a, b]$. We have the bounds

$$\max_{s \in [a,b]} \varphi(s) = \varphi\left(\frac{a + b}{2}\right) = 2^{1-r}(b - a)^r$$

and

$$\min_{s \in [a,b]} \varphi(s) = \varphi(a) = \varphi(b) = (b - a)^r.$$

This proves the last part of (3.7).

For $r = 1$ we have

$$B_1(a, b; s) := \sin\left(\frac{b - s}{2}\right) + \sin\left(\frac{s - a}{2}\right)$$

$$= 2\sin\left(\frac{b - a}{4}\right)\cos\left(\frac{s - \frac{a+b}{2}}{2}\right)$$

which implies that

$$\max_{s \in [a,b]} B_1(a, b; s) = 2\sin\left(\frac{b - a}{4}\right)\max_{s \in [a,b]}\cos\left(\frac{s - \frac{a+b}{2}}{2}\right)$$

$$= 2\sin\left(\frac{b - a}{4}\right) \leq \frac{b - a}{2}$$

which proves the desired result (3.9).

Now, for the best constant, assume that there is a $D > 0$ such that

$$\left| \frac{f\left(e^{ib}\right) + f\left(e^{ia}\right)}{2} \left[u\left(b\right) - u\left(a\right)\right] - \int_a^b f\left(e^{is}\right) du\left(s\right) \right| \quad (3.15)$$

$$\leq DK \sin\left(\frac{b-a}{4}\right) \bigvee_a^b \left(u\right)$$

for an interval $[a, b] \subseteq [0, 2\pi]$, a K-Lipschitzian function $f : \mathcal{C}\left(0, 1\right) \to \mathbb{C}$, and a function of bounded variation $u : [a, b] \to \mathbb{C}$.

If we take $[a, b] = [0, 2\pi]$, $f\left(z\right) = z$ then $K = 1$ and the inequality (3.15) becomes

$$\left| u\left(2\pi\right) - u\left(0\right) - \int_0^{2\pi} e^{is} du\left(s\right) \right| \leq D \bigvee_0^{2\pi} \left(u\right) \quad (3.16)$$

for any function of bounded variation $u : [0, 2\pi] \to \mathbb{C}$.

Integrating by parts in the Riemann–Stieltjes integral, we have

$$\int_0^{2\pi} e^{is} du\left(s\right) = e^{is} u\left(s\right) \Big|_0^{2\pi} - i \int_0^{2\pi} e^{is} u\left(s\right) ds$$

$$= u\left(2\pi\right) - u\left(0\right) - i \int_0^{2\pi} e^{is} u\left(s\right) ds$$

and the inequality (3.16) becomes

$$\left| \int_0^{2\pi} e^{is} u\left(s\right) ds \right| \leq D \bigvee_0^{2\pi} \left(u\right) \quad (3.17)$$

for any function of bounded variation $u : [0, 2\pi] \to \mathbb{C}$.

Now, if we take the function

$$u\left(s\right) := \begin{cases} -1 & \text{if } s \in [0, \pi] \\ \\ 1 & \text{if } s \in [\pi, 2\pi], \end{cases}$$

then u is of bounded variation, $\bigvee\limits_{0}^{2\pi}(u) = 2$ and

$$\int_0^{2\pi} e^{is} u(s)\, ds = -\int_0^\pi e^{is} ds + \int_\pi^{2\pi} e^{is} ds$$

$$= -\frac{1}{i}e^{i\pi} + \frac{1}{i}e^0 + \frac{1}{i}e^{2\pi} - \frac{1}{i}e^{i\pi} = \frac{4}{i}$$

and the inequality (3.17) becomes $4 \leq 2D$ showing that $D \geq 2$.

Remark 11 *If we take $a = 0$ and $b = 2\pi$, then we get from (3.9) that*

$$\left| f(1)\left[u(2\pi) - u(0)\right] - \int_0^{2\pi} f\left(e^{is}\right) du(s) \right| \leq 2K \bigvee_{0}^{2\pi}(u).$$
$$(3.18)$$

Remark 12 *If $0 < b - a \leq \pi$ then*

$$\max_{s\in[a,b]} B_r(a,b;s) \leq \max_{s\in[a,b]} \sin^r\left(\frac{b-s}{2}\right) + \max_{s\in[a,b]} \sin^r\left(\frac{s-a}{2}\right)$$

$$= 2\sin^r\left(\frac{b-a}{2}\right)$$

and by (3.7) we have

$$\left|T_C(f,u;a,b)\right| \leq 2^r H \sin^r\left(\frac{b-a}{2}\right) \bigvee_{a}^{b}(u).$$
$$(3.19)$$

Theorem 3.5 (Dragomir 2016, [21]) *Assume that $f : C(0,1) \rightarrow \mathbb{C}$ satisfies the Hölder-type condition (3.6). If $[a,b] \subseteq [0,2\pi]$ and the function $u : [a,b] \rightarrow \mathbb{C}$ is Lipschitzian with the constant $L > 0$ on $[a,b]$, then*

$$\left| \frac{f(e^{ib}) + f(e^{ia})}{2}\left[u(b) - u(a)\right] - \int_a^b f\left(e^{is}\right) du(s)\right| \quad (3.20)$$

$$\leq 2^{r-1} LH \int_a^b \left[\sin^r\left(\frac{b-s}{2}\right) + \sin^r\left(\frac{s-a}{2}\right)\right] ds$$

$$\leq LH \frac{(b-a)^{r+1}}{(r+1)}.$$

In particular, if $f : \mathcal{C}(0,1) \to \mathbb{C}$ is Lipschitzian with the constant $K > 0$, then we have

$$\left| \frac{f(e^{ib}) + f(e^{ia})}{2} [u(b) - u(a)] - \int_a^b f(e^{is}) \, du(s) \right| \quad (3.21)$$

$$\leq 8LK \sin^2 \left(\frac{b-a}{4} \right) \leq \frac{1}{2} LH (b-a)^2 .$$

Proof. It is well known that if $p : [a,b] \to \mathbb{C}$ is a Riemann integrable function and $v : [a,b] \to \mathbb{C}$ is Lipschitzian with the constant $M > 0$, then the Riemann–Stieltjes integral $\int_a^b p(t) \, dv(t)$ exists and the following inequality holds

$$\left| \int_a^b p(t) \, dv(t) \right| \leq M \int_a^b |p(t)| \, dt. \quad (3.22)$$

Taking the modulus in the equality (3.10) and utilizing the property (3.22) we deduce

$$\left| \frac{f(e^{ib}) + f(e^{ia})}{2} [u(b) - u(a)] - \int_a^b f(e^{is}) \, du(s) \right| \quad (3.23)$$

$$\leq \left| \int_a^b \left[\frac{f(e^{ib}) + f(e^{ia})}{2} - f(e^{is}) \right] du(s) \right|$$

$$\leq L \int_a^b \left| \frac{f(e^{ib}) + f(e^{ia})}{2} - f(e^{is}) \right| ds$$

$$\leq \frac{1}{2} L \int_a^b \left[\left| f(e^{ib}) - f(e^{is}) \right| + \left| f(e^{is}) - f(e^{ia}) \right| \right] ds$$

$$\leq \frac{1}{2} LH \int_a^b \left[\left| e^{ib} - e^{is} \right|^r + \left| e^{is} - e^{ia} \right|^r \right] ds$$

$$= 2^{r-1} LH \int_a^b \left[\sin^r \left(\frac{b-s}{2} \right) + \sin^r \left(\frac{s-a}{2} \right) \right] ds$$

which proves the first inequality in (3.20).

On making use of the elementary inequality $\sin x \leq x, x \in$

$[0, \pi]$ we have

$$
\int_a^b \left[\sin^r \left(\frac{b-s}{2} \right) + \sin^r \left(\frac{s-a}{2} \right) \right] ds
$$

$$
\leq \int_a^b \left(\frac{b-s}{2} \right)^r ds + \left(\frac{s-a}{2} \right)^r ds
$$

$$
= \frac{(b-a)^{r+1} + (b-a)^{r+1}}{(r+1)\, 2^r} = \frac{(b-a)^{r+1}}{(r+1)\, 2^{r-1}}.
$$

This proves the second part of the inequality (3.20).

For $r = 1$ we have

$$
\int_a^b \left[\sin \left(\frac{b-s}{2} \right) + \sin \left(\frac{s-a}{2} \right) \right] ds
$$

$$
= 2 \sin \left(\frac{b-a}{4} \right) \int_a^b \cos \left(\frac{s - \frac{a+b}{2}}{2} \right) ds = 8 \sin^2 \left(\frac{b-a}{4} \right).
$$

Using (3.20) for $r = 1$ we deduce (3.21).

Remark 13 *For $a = 0$ and $b = 2\pi$ we have by (3.21) that*

$$
\left| f(1) \left[u(2\pi) - u(0) \right] - \int_0^{2\pi} f\left(e^{is} \right) du(s) \right| \leq 8LK. \quad (3.24)
$$

The case of monotonic nondecreasing integrators that is important for applications for unitary operators is as follows.

Theorem 3.6 (Dragomir 2016, [21]) *Assume that $f :$ $\mathcal{C}(0,1) \to \mathbb{C}$ satisfies the Hölder-type condition (3.6). If $[a,b] \subseteq [0, 2\pi]$ and the function $u : [a,b] \to \mathbb{R}$ is monotonic nondecreasing on $[a,b]$, then*

$$
\left| \frac{f\left(e^{ib} \right) + f\left(e^{ia} \right)}{2} \left[u(b) - u(a) \right] - \int_a^b f\left(e^{is} \right) du(s) \right| \quad (3.25)
$$

$$
\leq 2^{r-1} H \int_a^b \left[\sin^r \left(\frac{b-s}{2} \right) + \sin^r \left(\frac{s-a}{2} \right) \right] du(s)
$$

$$
\leq \frac{1}{2} H \int_a^b \left[(b-s)^r + (s-a)^r \right] du(s).
$$

In particular, if $f : C(0,1) \to \mathbb{C}$ is Lipschitzian with the constant $K > 0$, then we have

$$\left| \frac{f(e^{ib}) + f(e^{ia})}{2} [u(b) - u(a)] - \int_a^b f(e^{is}) \, du(s) \right| \quad (3.26)$$

$$\leq 2^{1/2} K \sin\left(\frac{b-a}{4}\right) \int_a^b \left[1 + \cos\left(s - \frac{a+b}{2}\right) \right]^{1/2} du(s)$$

$$\leq \frac{1}{2} K(b-a)[u(b) - u(a)].$$

Proof. It is well known that if $p : [a,b] \to \mathbb{C}$ is a continuous function and $v : [a,b] \to \mathbb{R}$ is monotonic nondecreasing on $[a,b]$, then the Riemann–Stieltjes integral $\int_a^b p(t) \, dv(t)$ exists and the following inequality holds

$$\left| \int_a^b p(t) \, dv(t) \right| \leq \int_a^b |p(t)| \, dv(t). \quad (3.27)$$

Utilizing the property (3.27), we have from (3.10) that

$$\left| \frac{f(e^{ib}) + f(e^{ia})}{2} [u(b) - u(a)] - \int_a^b f(e^{is}) \, du(s) \right| \quad (3.28)$$

$$\leq \left| \int_a^b \left[\frac{f(e^{ib}) + f(e^{ia})}{2} - f(e^{is}) \right] du(s) \right|$$

$$\leq \int_a^b \left| \frac{f(e^{ib}) + f(e^{ia})}{2} - f(e^{is}) \right| du(s)$$

$$\leq \frac{1}{2} \int_a^b \left[\left| f(e^{ib}) - f(e^{is}) \right| + \left| f(e^{is}) - f(e^{ia}) \right| \right] du(s)$$

$$\leq \frac{1}{2} H \int_a^b \left[\left| e^{ib} - e^{is} \right|^r + \left| e^{is} - e^{ia} \right|^r \right] du(s)$$

$$= 2^{r-1} H \int_a^b \left[\sin^r \left(\frac{b-s}{2} \right) + \sin^r \left(\frac{s-a}{2} \right) \right] du(s),$$

which proves the first part of (3.25). The second part is obvious.

For $r = 1$ we have

$$\int_a^b \left[\sin \left(\frac{b-s}{2} \right) + \sin \left(\frac{s-a}{2} \right) \right] du(s)$$

$$= 2 \sin \left(\frac{b-a}{4} \right) \int_a^b \cos \left(\frac{s - \frac{a+b}{2}}{2} \right) u(s)$$

$$= 2^{1/2} \sin \left(\frac{b-a}{4} \right) \int_a^b \left[1 + \cos \left(s - \frac{a+b}{2} \right) \right]^{1/2} du(s).$$

This proves (3.26).

Corollary 3.7 (Dragomir 2016, [21]) *Assume that f is as in Theorem 3.6. If the function $u : [0, 2\pi] \to \mathbb{R}$ is monotonic nondecreasing on $[0, 2\pi]$, then*

$$\left| f(1) [u(2\pi) - u(0)] - \int_0^{2\pi} f(e^{is}) du(s) \right| \qquad (3.29)$$

$$\leq 2^r H \int_0^{2\pi} \sin^r \left(\frac{s}{2} \right) du(s) = 2^{r/2} H \int_0^{2\pi} (1 - \cos s)^{r/2} du(s).$$

Proof. We have

$$\int_0^{2\pi} \left[\sin^r \left(\frac{2\pi - s}{2} \right) + \sin^r \left(\frac{s}{2} \right) \right] du(s)$$

$$= \int_0^{2\pi} \left[\sin^r \left(\pi - \frac{s}{2} \right) + \sin^r \left(\frac{s}{2} \right) \right] du(s)$$

$$= \int_t^{2\pi} \left[\sin^r \left(\frac{s}{2} \right) + \sin^r \left(\frac{s}{2} \right) \right] du(s)$$

$$= 2 \int_0^{2\pi} \sin^r \left(\frac{s}{2} \right) du(s)$$

and by (3.25) we get (3.29).

Since for $s \in [0, 2\pi]$ we have

$$\sin \left(\frac{s}{2} \right) = \left(\frac{1 - \cos s}{2} \right)^{1/2},$$

then the last part of (3.29) is obtained.

3.3 A QUADRATURE RULE

We consider the following *partition* of the interval $[a, b]$

$$\Delta_n : a = x_0 < x_1 < ... < x_{n-1} < x_n = b$$

where $0 \le k \le n-1$. Define $h_k := x_{k+1} - x_k$, $0 \le k \le n-1$ and $\nu(\Delta_n) = \max\{h_k : 0 \le k \le n-1\}$ the norm of the partition Δ_n.

For the continuous function $f : \mathcal{C}(0, 1) \to \mathbb{C}$ and the function $u : [a, b] \subseteq [0, 2\pi] \to \mathbb{C}$ of bounded variation on $[a, b]$, define the *trapezoid quadrature rule*

$$T_n(f, u, \Delta_n) := \sum_{k=0}^{n-1} \frac{f\left(e^{ix_{k+1}}\right) + f\left(e^{ix_k}\right)}{2} \left[u(x_{k+1}) - u(x_k)\right]$$

(3.30)

and the remainder $R_n(f, u, \Delta_n)$ in approximating the Riemann–Stieltjes integral $\int_a^b f\left(e^{it}\right) du(t)$ by $T_n(f, u, \Delta_n)$. Then we have

$$\int_a^b f\left(e^{it}\right) du(t) = T_n(f, u, \Delta_n) + R_n(f, u, \Delta_n).$$ (3.31)

The following result provides *a priori* bounds for $R_n(f, u, \Delta_n)$ in several instances of f and u as above.

Proposition 3.8 (Dragomir 2016, [21]) *Assume that $f : \mathcal{C}(0, 1) \to \mathbb{C}$ satisfies the following Hölder-type condition*

$$|f(z) - f(w)| \le H |z - w|^r$$

for any $w, z \in \mathcal{C}(0, 1)$, where $H > 0$ and $r \in (0, 1]$ are given.

If $[a, b] \subseteq [0, 2\pi]$ and the function $u : [a, b] \to \mathbb{C}$ is of bounded variation on $[a, b]$, then for any partition $\Delta_n : a = x_0 < x_1 < ... < x_{n-1} < x_n = b$ with the norm $\nu(\Delta_n) \le \pi$ we have the error bound

$$|R_n(f, u, \Delta_n)| \le 2^r H \sum_{k=0}^{n-1} \sin^r\left(\frac{x_{k+1} - x_k}{2}\right) \bigvee_{x_k}^{x_{k+1}}(u)$$ (3.32)

$$\le 2^r H \sin^r\left(\frac{\nu(\Delta_n)}{2}\right) \bigvee_a^b(u) \le \nu^r(\Delta_n) H \bigvee_a^b(u).$$

Proof. Since $\nu(\Delta_n) \leq \pi$, then on writing inequality (3.19) on each interval $[x_k, x_{k+1}]$ and for any intermediate points $\xi_k \in [x_k, x_{k+1}]$ where $0 \leq k \leq n-1$, we have

$$\left| \int_{x_k}^{x_{k+1}} f\left(e^{it}\right) du(t) - \frac{f\left(e^{ix_{k+1}}\right) + f\left(e^{ix_k}\right)}{2} \left[u(x_{k+1}) - u(x_k) \right] \right|$$

$$(3.33)$$

$$\leq 2^r H \sin^r \left(\frac{x_{k+1} - x_k}{2} \right) \bigvee_{x_k}^{x_{k+1}} (u) \leq 2^r H \sin^r \left(\frac{\nu(\Delta_n)}{2} \right) \bigvee_{x_k}^{x_{k+1}} (u)$$

$$\leq \nu^r(\Delta_n) H \bigvee_{x_k}^{x_{k+1}} (u).$$

Summing over k from 0 to $n-1$ in (3.33) and utilizing the generalized triangle inequality, we deduce (3.32).

Remark 14 *If the function $f : \mathcal{C}(0,1) \to \mathbb{C}$ is Lipschitzian with the constant $K > 0$, then by (3.9) we have a better error bound, namely*

$$|R_n(f, u, \Delta_n)| \leq 2K \sum_{k=0}^{n-1} \sin \left(\frac{x_{k+1} - x_k}{4} \right) \bigvee_{x_k}^{x_{k+1}} (u) \qquad (3.34)$$

$$\leq 2K \sin \left(\frac{\nu(\Delta_n)}{4} \right) \bigvee_{a}^{b} (u) \leq \frac{1}{2} \nu(\Delta_n) K \bigvee_{a}^{b} (u).$$

Remark 15 *The inequality (3.34) has some particular cases of interest as follows.*

1. If we take $\Delta_2 : a = x_0 = 0$, $x_1 = \pi$, $x_2 = b = 2\pi$, then

$$T_2(f, u, \Delta_2) = \frac{f(-1) + f(1)}{2} \left[u(\pi) - u(0) \right]$$

$$+ \frac{f(1) + f(-1)}{2} \left[u(2\pi) - u(\pi) \right]$$

$$= \frac{f(1) + f(-1)}{2} \left[u(2\pi) - u(0) \right]$$

and writing the inequality (3.34) for this case we get

$$\left| \int_0^{2\pi} f\left(e^{it}\right) du(t) - \frac{f(1) + f(-1)}{2} \left[u(2\pi) - u(0)\right] \right|$$

$$\leq \sqrt{2} K \bigvee_0^{2\pi} (u).$$ (3.35)

2. If we take $\Delta_4 : a = x_0 = 0$, $x_1 = \frac{\pi}{2}$, $x_2 = \pi$, $x_3 = \frac{3\pi}{2}$, $x_4 = b = 2\pi$, *then*

$$T_4(f, u, \Delta_4)$$
$$= \frac{f(i) + f(1)}{2} \left[u\left(\frac{\pi}{2}\right) - u(0)\right]$$
$$+ \frac{f(i) + f(-1)}{2} \left[u(\pi) - u\left(\frac{\pi}{2}\right)\right]$$
$$+ \frac{f(-1) + f(-i)}{2} \left[u\left(\frac{3\pi}{2}\right) - u(\pi)\right]$$
$$+ \frac{f(-i) + f(1)}{2} \left[u(2\pi) - u\left(\frac{3\pi}{2}\right)\right]$$
$$= \frac{f(1)}{2} \left[u(2\pi) - u\left(\frac{3\pi}{2}\right) + u\left(\frac{\pi}{2}\right) - u(0)\right]$$
$$+ \frac{f(i)}{2} \left[u(\pi) - u(0)\right]$$
$$+ \frac{f(-1)}{2} \left[u\left(\frac{3\pi}{2}\right) - u\left(\frac{\pi}{2}\right)\right] + \frac{f(-i)}{2} \left[u(2\pi) - u(\pi)\right]$$

and writing the inequality (3.34) for this case we get

$$\left| \int_0^{2\pi} f\left(e^{it}\right) du(t) - \frac{f(1)}{2} \left[u(2\pi) - u\left(\frac{3\pi}{2}\right) + u\left(\frac{\pi}{2}\right) - u(0)\right] \right.$$ (3.36)

$$- \frac{f(i)}{2} \left[u(\pi) - u(0)\right] - \frac{f(-1)}{2} \left[u\left(\frac{3\pi}{2}\right) - u\left(\frac{\pi}{2}\right)\right]$$

$$\left. - \frac{f(-i)}{2} \left[u(2\pi) - u(\pi)\right] \right|$$

$$\leq \sqrt{2 - \sqrt{2}} K \bigvee_0^{2\pi} (u).$$

3.4 APPLICATIONS FOR UNITARY OPERATORS

We consider the following partition of the interval $[0, 2\pi]$

$$\Gamma_n : 0 = \lambda_0 < \lambda_1 < \dots < \lambda_{n-1} < \lambda_n = 2\pi$$

and the intermediate points $\xi_k \in [\lambda_k, \lambda_{k+1}]$ where $0 \leq k \leq n - 1$. Define $h_k := \lambda_{k+1} - \lambda_k$, $0 \leq k \leq n - 1$ and $\nu(\Gamma_n) = \max\{h_k : 0 \leq k \leq n - 1\}$ the norm of the partition Γ_n.

If U is a unitary operator on the Hilbert space H and $\{E_\lambda\}_{\lambda \in [0,2\pi]}$, the spectral family of U, then we can introduce the following sums

$$T_n(f, u, \Gamma_n; x, y) := \sum_{k=0}^{n-1} \frac{f\left(e^{ix_{k+1}}\right) + f\left(e^{ix_k}\right)}{2}$$
$$\left\langle \left(E_{\lambda_{k+1}} - E_{\lambda_k}\right) x, y \right\rangle \qquad (3.37)$$

for $x, y \in H$.

For a function $f : \mathcal{C}(0, 1) \to \mathbb{C}$ that satisfies a Lipschitz-type condition with a constant $K > 0$, we can approximate the function f of unitary operator U as follows

$$\langle f(U) x, y \rangle = T_n(f, \Gamma_n; x, y) + R_n(f, \Gamma_n; x, y) \qquad (3.38)$$

for $x, y \in H$, where the reminder satisfies the bounds

$$|R_n(f, \Gamma_n; x, y)| \leq 2K \sum_{k=0}^{n-1} \sin\left(\frac{\lambda_{k+1} - \lambda_k}{4}\right) \bigvee_{\lambda_k}^{\lambda_{k+1}} \left(\left\langle E_{(\cdot)} x, y \right\rangle\right)$$
$$(3.39)$$

$$\leq 2K \sin\left(\frac{\nu(\Gamma_n)}{4}\right) \bigvee_{0}^{2\pi} \left(\left\langle E_{(\cdot)} x, y \right\rangle\right)$$

$$\leq \frac{1}{2} \nu(\Gamma_n) K \bigvee_{0}^{2\pi} \left(\left\langle E_{(\cdot)} x, y \right\rangle\right)$$

for any $x, y \in H$.

Since the following *Total Variation Schwarz-type inequality* holds (for a short proof see for instance [19]):

$$\bigvee_{0}^{2\pi} \left(\left\langle E_{(\cdot)} x, y \right\rangle\right) \leq \|x\| \|y\| \qquad (3.40)$$

for any $x, y \in H$, then in the bounds above we can replace $\bigvee\limits_{0}^{2\pi} \left(\left\langle E_{(\cdot)} x, y \right\rangle \right)$ with $\|x\| \|y\|$.

From (3.35) we have the following trapezoid-type inequality for K-Lipschitzian functions $f : \mathcal{C}(0,1) \to \mathbb{C}$ of unitary operators U

$$\left| \left\langle f(U) x, y \right\rangle - \frac{f(1) + f(-1)}{2} \left\langle x, y \right\rangle \right| \leq \sqrt{2} K \bigvee\limits_{0}^{2\pi} \left(\left\langle E_{(\cdot)} x, y \right\rangle \right)$$

(3.41)

$$\leq \sqrt{2} K \|x\| \|y\|$$

for any $x, y \in H$.

For $a \neq \pm 1, 0$ consider the function $f : \mathcal{C}(0,1) \to \mathbb{C}$, $f_a(z) = \frac{1}{1-az}$. Observe that

$$|f_a(z) - f_a(w)| = \frac{|a| |z - w|}{|1 - az| |1 - aw|}$$

(3.42)

for any $z, w \in \mathcal{C}(0,1)$.

If $z = e^{it}$ with $t \in [0, 2\pi]$, then we have

$$|1 - az|^2 = 1 - 2a \operatorname{Re}(\bar{z}) + a^2 |z|^2 = 1 - 2a \cos t + a^2$$
$$\geq 1 - 2|a| + a^2 = (1 - |a|)^2$$

therefore

$$\frac{1}{|1 - az|} \leq \frac{1}{|1 - |a||} \quad \text{and} \quad \frac{1}{|1 - aw|} \leq \frac{1}{|1 - |a||}$$

(3.43)

for any $z, w \in \mathcal{C}(0,1)$.

Utilizing (3.42) and (3.43) we deduce

$$|f_a(z) - f_a(w)| \leq \frac{|a|}{(1 - |a|)^2} |z - w|$$

(3.44)

for any $z, w \in \mathcal{C}(0,1)$, showing that the function f_a is Lipschitzian with the constant $L_a = \frac{|a|}{(1-|a|)^2}$ on the circle $\mathcal{C}(0,1)$.

If we write the inequality (3.41) for the function f_a, we get

$$\left| \left\langle (1 - aU)^{-1} x, y \right\rangle - \frac{1}{1 - a^2} \left\langle x, y \right\rangle \right| \leq \frac{\sqrt{2} \, |a|}{(1 - |a|)^2} \bigvee_0^{2\pi} \left(\left\langle E_{(\cdot)} x, y \right\rangle \right)$$

(3.45)

$$\leq \frac{\sqrt{2} \, |a|}{(1 - |a|)^2} K \, \|x\| \, \|y\|$$

for any $x, y \in H$.

Now, for $z, w \in \mathbb{C}$ define the function $f_{z,w} : [0, 1] \to \mathbb{C}$, $f_{z,w}(t) = \exp\left[(1 - t) z + tw\right]$. We observe that $f_{z,w}$ is differentiable on $(0, 1)$ and

$$\frac{df_{z,w}(t)}{dt} = (w - z) \exp\left[(1 - t) z + tw\right]$$

for any $t \in (0, 1)$.

We then have

$$|\exp(w) - \exp(z)| = |f_{z,w}(1) - f_{z,w}(0)| = \left| \int_0^1 \frac{df_{z,w}(t)}{dt} \, dt \right|$$

(3.46)

$$= \left| (w - z) \int_0^1 \exp\left[(1 - t) z + tw\right] dt \right|$$

$$\leq |w - z| \int_0^1 |\exp\left[(1 - t) z + tw\right]| \, dt$$

$$\leq |w - z| \int_0^1 \exp |(1 - t) z + tw| \, dt$$

$$\leq |w - z| \int_0^1 \exp\left[(1 - t) |z| + t |w|\right] dt$$

for any $z, w \in \mathbb{C}$. To obtain this we used the well-known inequality $|\exp(u)| \leq \exp(|u|)$ for any $u \in \mathbb{C}$.

We observe that if $u \in \mathbb{C}$, then

$$\begin{aligned} |\exp(u)| &= |\exp(\operatorname{Re} u + i \operatorname{Im} u)| = |\exp(\operatorname{Re} u)| \, |\exp(i \operatorname{Im} u)| \\ &= \exp(\operatorname{Re} u) \, |\cos(\operatorname{Im} u) + i \sin(\operatorname{Im} u)| = \exp(\operatorname{Re} u). \end{aligned}$$

Therefore

$$|\exp\left[(1-t)z+tw\right]| = \exp\left(\operatorname{Re}\left[(1-t)z+tw\right]\right)$$
$$= \exp\left[(1-t)\operatorname{Re}z+t\operatorname{Re}w\right]$$

for any $t \in [0,1]$.

From this inequality, we deduce the following result of interest

$$|\exp\left(w\right)-\exp\left(z\right)| \leq |w-z|\int_0^1 \exp\left[(1-t)\operatorname{Re}z+t\operatorname{Re}w\right]dt \tag{3.47}$$

that holds for any $z, w \in \mathbb{C}$.

In the case when $\operatorname{Re}z \neq \operatorname{Re}w$ we have

$$\int_0^1 \exp\left[(1-t)\operatorname{Re}z+t\operatorname{Re}w\right]dt = \frac{\exp\left(\operatorname{Re}z\right)-\exp\left(\operatorname{Re}w\right)}{\operatorname{Re}z-\operatorname{Re}w},$$

which implies the following inequality of interest:

$$\left|\frac{\exp\left(w\right)-\exp\left(z\right)}{w-z}\right| \leq \frac{\exp\left(\operatorname{Re}z\right)-\exp\left(\operatorname{Re}w\right)}{\operatorname{Re}z-\operatorname{Re}w}$$

that holds for any $z, w \in \mathbb{C}$ with $\operatorname{Re}z \neq \operatorname{Re}w$.

Now, if $w \in \mathbb{C}$ with $|w| = |z| = 1$, then from (3.46) we have

$$|\exp\left(w\right)-\exp\left(z\right)| \leq e|w-z|$$

which shows that the function $f(z) = \exp(z)$ is Lipschitzian with the constant $L = e$ on the circle $\mathcal{C}(0,1)$.

Utilizing the inequality (3.41) we have for any unitary operators U

$$\left|\langle \exp\left(U\right)x,y\rangle - \frac{e^2+1}{2e}\langle x,y\rangle\right| \leq \sqrt{2}e\bigvee_0^{2\pi}\left(\langle E_{(\cdot)}x,y\rangle\right) \tag{3.48}$$
$$\leq \sqrt{2}e\,\|x\|\,\|y\|$$

for any $x, y \in H$.

The interested reader may apply the above results for other Lipschitzian functions. However, the details are not presented here.

CHAPTER 4

Generalized Trapezoid Inequalities

I N THIS CHAPTER we present some results in approximating the Riemann–Stieltjes integral by generalized trapezoidal rule for continuous complex-valued integrands and various classes of bounded variation integrators. Some applications for functions of unitary operators in Hilbert spaces are provided as well.

4.1 INTRODUCTION

In [24], in order to approximate the *Riemann–Stieltjes integral* $\int_a^b f(t)\, du(t)$ by the *generalized trapezoid formula*

$$[u(b) - u(x)] f(b) + [u(x) - u(a)] f(a), \qquad x \in [a, b] \quad (4.1)$$

the authors considered the error functional

$$T(f, u; a, b; x)$$
$$:= \int_a^b f(t)\, du(t) - [u(b) - u(x)] f(b) - [u(x) - u(a)] f(a)$$
$$(4.2)$$

and proved that

$$|T(f, u; a, b; x)| \leq H \left[\frac{1}{2}(b-a) + \left| x - \frac{a+b}{2} \right| \right]^r \bigvee_a^b (f), \quad (4.3)$$

for any $x \in [a, b]$, provided that $f : [a, b] \to \mathbb{R}$ is of *bounded variation* on $[a, b]$ and u is of $r - H$-*Hölder type*, that is, $u : [a, b] \to \mathbb{R}$ satisfies the condition $|u(t) - u(s)| \leq H |t - s|^r$ for any $t, s \in [a, b]$, where $r \in (0, 1]$ and $H > 0$ are given.

The dual case, namely, when f is of $q - K$-Hölder type and u is of bounded variation has been considered by the authors in [4] in which they obtained the bound:

$$|T(f, u; a, b; x)| \quad (4.4)$$

$$\leq K \left[(x-a)^q \bigvee_a^x (u) + (b-x)^q \bigvee_x^b (u) \right]$$

$$\leq \begin{cases} K \left[(x-a)^q + (b-x)^q \right] \left[\frac{1}{2} V_a^b(u) + \frac{1}{2} \left| V_a^x(u) - V_x^b(u) \right| \right]; \\[2ex] K \left[(x-a)^{q\alpha} + (b-x)^{q\alpha} \right]^{\frac{1}{\alpha}} \left[[V_a^x(u)]^\beta - \left[V_x^b(u) \right]^\beta \right]^{\frac{1}{\beta}} \\[1ex] \qquad \qquad \text{if } \alpha > 1, \ \frac{1}{\alpha} + \frac{1}{\beta} = 1; \\[2ex] K \left[\frac{1}{2}(b-a) + \left| x - \frac{a+b}{2} \right| \right]^q V_a^b(u), \end{cases}$$

for any $x \in [a, b]$.

The case where f is monotonic and u is of $r - H$-Hölder type, which provides a refinement for (4.3), respectively the case where u is monotonic and f of $q - K$-Hölder type were considered by Cheung and Dragomir in [6], while the case where one function was of Hölder type and the other was Lipschitzian were considered in [2]. For other recent results in estimating the error $T(f, u; a, b, x)$ for absolutely continuous integrands f and integrators u of bounded variation, see [5] and [3].

Motivated by the above facts, we consider in this chapter the problem of approximating the Riemann–Stieltjes integral $\int_a^b f(e^{is}) \, du(s)$ by the *generalized trapezoidal rule*

$$f\left(e^{ib}\right) [u(b) - u(t)] + f\left(e^{ia}\right) [u(t) - u(a)]$$

for continuous complex-valued function $f : \mathcal{C}(0,1) \to \mathbb{C}$ defined on the complex unit circle $\mathcal{C}(0,1)$ and various subclasses of functions $u : [a,b] \subseteq [0,2\pi] \to \mathbb{C}$ of bounded variation.

We denote the *generalized trapezoidal error functional* by

$$T_{\mathcal{C}}(f,u;a,b;t) \qquad\qquad (4.5)$$
$$:= f\left(e^{ib}\right)[u(b) - u(t)] + f\left(e^{ia}\right)[u(t) - u(a)]$$
$$- \int_a^b f\left(e^{is}\right) du(s),$$

where $t \in [a,b]$ and will provide some bounds for its magnitude for f of $r - H-$Hölder type and u belonging to different subclasses of functions of bounded variation.

The Riemann–Stieltjes integral $\int_0^{2\pi} f\left(e^{is}\right) du(s)$ is related to functions of unitary operators U defined on complex Hilbert spaces as follows.

4.2 GENERALIZED TRAPEZOID INEQUALITIES

We have the following result.

Theorem 4.1 (Dragomir 2015, [17]) *Assume that* $f : \mathcal{C}(0,1) \to \mathbb{C}$ *satisfies the following Hölder-type condition*

$$|f(z) - f(w)| \le H|z - w|^r \qquad\qquad (4.6)$$

for any $w, z \in \mathcal{C}(0,1),$ *where* $H > 0$ *and* $r \in (0,1]$ *are given.*

If $[a,b] \subseteq [0,2\pi]$ *and the function* $u : [a,b] \to \mathbb{C}$ *is of bounded variation on* $[a,b],$ *then*

$$|T_{\mathcal{C}}(f,u;a,b;t)| \le 2^r H B_r(u;a,b,t) \qquad\qquad (4.7)$$

for any $t \in [a,b],$ *where the bound* $B_r(u;a,b;t)$ *is given by*

$$B_r(u;a,b;t) \qquad\qquad (4.8)$$
$$:= \max_{s\in[a,t]}\left\{\sin^r\left(\frac{s-a}{2}\right)\right\} \bigvee_a^t(u) + \max_{s\in[t,b]}\left\{\sin^r\left(\frac{b-s}{2}\right)\right\} \bigvee_t^b(u).$$

Moreover, if we denote

$$A_r(t) := \max_{s \in [a,t]} \left\{ \sin^r \left(\frac{s-a}{2} \right) \right\} \text{ and } B_r(t)$$

$$:= \max_{s \in [t,b]} \left\{ \sin^r \left(\frac{b-s}{2} \right) \right\},$$

then we have the inequalities:

$$B_r(u; a, b; t) \tag{4.9}$$

$$\leq \begin{cases} \max\{A_r(t), B_r(t)\} \bigvee_a^b (u) \\ \\ [A_r^\alpha(t) + B_r^\alpha(t)]^{\frac{1}{\alpha}} \left[\left(\bigvee_a^t (u) \right)^\beta + \left(\bigvee_t^b (u) \right)^\beta \right]^{\frac{1}{\beta}} \\ \alpha > 1, \frac{1}{\alpha} + \frac{1}{\beta} = 1 \\ \\ [A_r(t) + B_r(t)] \max \left\{ \bigvee_a^t (u), \bigvee_t^b (u) \right\} \end{cases}$$

for any $t \in [a, b]$.

Proof. We have the equality

$$T_C(f, u; a, b; t) \tag{4.10}$$

$$= \int_t^b \left[f\left(e^{ib}\right) - f\left(e^{is}\right) \right] du(s) + \int_a^t \left[f\left(e^{is}\right) - f\left(e^{ia}\right) \right] du(s)$$

for any $t \in [a, b]$.

It is known that if $p : [c, d] \to \mathbb{C}$ is a continuous function and $v : [c, d] \to \mathbb{C}$ is of bounded variation, then the Riemann–Stieltjes integral $\int_c^d p(t) \, dv(t)$ exists and the following inequality holds

$$\left| \int_c^d p(t) \, dv(t) \right| \leq \max_{t \in [c,d]} |p(t)| \bigvee_c^d (v). \tag{4.11}$$

Taking the modulus in the equality (4.10) and utilizing the property (4.11) we deduce

$$|T_{\mathcal{C}}(f, u; a, b; t)| \tag{4.12}$$

$$\leq \left| \int_t^b \left[f\left(e^{ib}\right) - f\left(e^{is}\right) \right] du(s) \right|$$

$$+ \left| \int_a^t \left[f\left(e^{is}\right) - f\left(e^{ia}\right) \right] du(s) \right|$$

$$\leq \max_{s \in [t,b]} \left| f\left(e^{ib}\right) - f\left(c^{is}\right) \right| \bigvee_t^b (u)$$

$$+ \max_{s \in [a,t]} \left| f\left(e^{is}\right) - f\left(e^{ia}\right) \right| \bigvee_a^t (u)$$

$$\leq H \left[\max_{s \in [t,b]} \left| e^{ib} - e^{is} \right|^r \bigvee_t^b (u) + \max_{s \in [a,t]} \left| e^{is} - e^{ia} \right|^r \bigvee_a^t (u) \right]$$

for any $t \in [a, b]$.

Since

$$\left| e^{is} - e^{it} \right|^2 = \left| e^{is} \right|^2 - 2 \operatorname{Re}\left(e^{i(s-t)} \right) + \left| e^{it} \right|^2$$

$$= 2 - 2\cos(s-t) = 4\sin^2\left(\frac{s-t}{2} \right)$$

for any $t, s \in \mathbb{R}$, then

$$\left| e^{is} - e^{it} \right|^r = 2^r \left| \sin\left(\frac{s-t}{2} \right) \right|^r \tag{4.13}$$

for any $t, s \in \mathbb{R}$.

For $[a, b] \subseteq [0, 2\pi]$ we have

$$\left| e^{ib} - e^{is} \right|^r = 2^r \sin^r \left(\frac{b-s}{2} \right)$$

and

$$\left| e^{is} - e^{ia} \right|^r = 2^r \sin^r \left(\frac{s-a}{2} \right)$$

for any $s \in [a, b]$.

Utilizing the inequality (4.12) we deduce the desired result (4.7).

By making use of the Hölder inequality

$$
mp + nq \leq
\begin{cases}
\max\{m, n\}\, (p + q) \\[2mm]
\left(m^{1/\alpha} + n^{1/\alpha}\right)^{\alpha} \left(p^{1/\beta} + q^{1/\beta}\right), \\[1mm]
\quad \alpha > 1,\ \frac{1}{\alpha} + \frac{1}{\beta} = 1
\end{cases}
$$

that holds for $m, p, n, q \geq 0$, we deduce the inequality (4.9).

Define the functional

$$
M_{\mathcal{C}}\left(f, u; a, b\right) := T_{\mathcal{C}}\left(f, u; a, b; \frac{a+b}{2}\right) \tag{4.14}
$$

$$
= f\left(e^{ib}\right)\left[u\left(b\right) - u\left(\frac{a+b}{2}\right)\right]
$$

$$
+ f\left(e^{ia}\right)\left[u\left(\frac{a+b}{2}\right) - u\left(a\right)\right]
$$

$$
- \int_{a}^{b} f\left(e^{is}\right) du\left(s\right).
$$

For $t = \frac{a+b}{2}$ we have

$$
A_r\left(\frac{a+b}{2}\right) := \max_{s \in \left[a, \frac{a+b}{2}\right]} \left\{\sin^r\left(\frac{s-a}{2}\right)\right\} = \sin^r\left(\frac{b-a}{4}\right),
$$

and

$$
B_r\left(\frac{a+b}{2}\right) := \max_{s \in \left[\frac{a+b}{2}, b\right]} \left\{\sin^r\left(\frac{b-s}{2}\right)\right\} = \sin^r\left(\frac{b-a}{4}\right).
$$

We can then state the following particular case of interest for applications as shown below.

Corollary 4.2 (Dragomir 2015, [17]) *With the assumptions of Theorem 4.1 we have*

$$
\left|M_{\mathcal{C}}\left(f, u; a, b\right)\right| \leq 2^r H \sin^r\left(\frac{b-a}{4}\right) \bigvee_{a}^{b}\left(u\right) \tag{4.15}
$$

$$
\leq \frac{1}{2^r} H\left(b-a\right)^r \bigvee_{a}^{b}\left(u\right).
$$

In particular, if f is Lipschitzian with the constant $K > 0$, then

$$|M_C(f, u; a, b)| \leq 2K \sin\left(\frac{b-a}{4}\right) \bigvee_a^b (u) \leq \frac{1}{2}K(b-a) \bigvee_a^b (u).$$

(4.16)

The constant 2 in the first inequality (4.16) is the best possible in the sense that it cannot be replaced by a smaller quantity.

Proof. We must only prove the sharpness of the constant 2 in the first inequality (4.16).

Assume that there is an $E > 0$ such that

$$\left| f\left(e^{ib}\right)\left[u(b) - u\left(\frac{a+b}{2}\right)\right] + f\left(e^{ia}\right)\left[u\left(\frac{a+b}{2}\right) - u(a)\right] \right.$$

(4.17)

$$\left. - \int_a^b f\left(e^{is}\right) du(s) \right|$$

$$\leq EK \sin\left(\frac{b-a}{4}\right) \bigvee_a^b (u)$$

for an interval $[a, b] \subseteq [0, 2\pi]$, a K-Lipschitzian function $f : C(0, 1) \to \mathbb{C}$ and a function of bounded variation $u : [a, b] \to \mathbb{C}$.

If we take $[a, b] = [0, 2\pi]$, $f(z) - z$ then $K = 1$ and the inequality (4.17) becomes

$$\left| u(2\pi) - u(0) - \int_0^{2\pi} e^{is} du(s) \right| \leq E \bigvee_0^{2\pi} (u)$$

(4.18)

for any function of bounded variation $u : [0, 2\pi] \to \mathbb{C}$.

Integrating by parts in the Riemann–Stieltjes integral, we have

$$\int_0^{2\pi} e^{is} du(s) = e^{is} u(s) \Big|_0^{2\pi} - i \int_0^{2\pi} e^{is} u(s) \, ds$$

$$= u(2\pi) - u(0) - i \int_0^{2\pi} e^{is} u(s) \, ds$$

and the inequality (4.18) becomes

$$\left| \int_0^{2\pi} e^{is} u(s)\, ds \right| \le E \bigvee_0^{2\pi} (u) \tag{4.19}$$

for any function of bounded variation $u : [0, 2\pi] \to \mathbb{C}$.

Now, if we take the function

$$u(s) := \begin{cases} -1 & \text{if } s \in [0, \pi] \\[2mm] 1 & \text{if } s \in [\pi, 2\pi], \end{cases}$$

then u is of bounded variation, $\bigvee_0^{2\pi} (u) = 2$ and

$$\int_0^{2\pi} e^{is} u(s)\, ds = - \int_0^\pi e^{is} ds + \int_\pi^{2\pi} e^{is} ds$$

$$= -\frac{1}{i} e^{i\pi} + \frac{1}{i} e^0 + \frac{1}{i} e^{2\pi} - \frac{1}{i} e^{i\pi} = \frac{4}{i}$$

and the inequality (4.19) becomes $4 \le 2E$ showing that $E \ge 2$.

Remark 16 *If the length of the interval $[a, b]$ is less than π, i.e., $0 < b - a \le \pi$, then*

$$A_r(t) := \sin^r \left(\frac{t - a}{2} \right) \quad \text{and} \quad B_r(t) := \sin^r \left(\frac{b - t}{2} \right)$$

and by (4.7) and by (4.9) we have

$$|T_{\mathcal{C}}(f,u;a,b;t)| \tag{4.20}$$

$$\leq 2^r H \left[\sin^r \left(\frac{t-a}{2} \right) \bigvee_a^t (u) + \sin^r \left(\frac{b-t}{2} \right) \bigvee_t^b (u) \right]$$

$$\leq \begin{cases} \max \left\{ \sin^r \left(\frac{t-a}{2} \right), \sin^r \left(\frac{b-t}{2} \right) \right\} \bigvee_a^b (u) \\ \\ \left[\sin^{\alpha r} \left(\frac{t-a}{2} \right) + \sin^{\alpha r} \left(\frac{b-t}{2} \right) \right]^{\frac{1}{\alpha}} \\ \times \left[\left(\bigvee_a^t (u) \right)^{\beta} + \left(\bigvee_t^b (u) \right)^{\beta} \right]^{\frac{1}{\beta}} \qquad \begin{array}{l} \alpha > 1, \\ \frac{1}{\alpha} + \frac{1}{\beta} = 1 \end{array} \\ \\ \left[\sin^r \left(\frac{t-a}{2} \right) + \sin^r \left(\frac{b-t}{2} \right) \right] \max \left\{ \bigvee_a^t (u), \bigvee_t^b (u) \right\} \end{cases}$$

for any $t \in [a,b]$.

Remark 17 *If* $a = 0$ *and* $b = 2\pi$, *then*

$$T_{\mathcal{C}}(f,u;0,2\pi;\pi) = f(1)[u(2\pi) - u(0)] - \int_0^{2\pi} f\left(e^{is}\right) du(s),$$

and

$$A_r(\pi) := \max_{s \in [0,\pi]} \left\{ \sin^r \left(\frac{s}{2} \right) \right\} = 1$$

while

$$B_r(\pi) := \max_{s \in [\pi,2\pi]} \left\{ \sin^r \left(\frac{2\pi - s}{2} \right) \right\} = 1.$$

Therefore from (4.7) we have

$$\left| f(1)[u(2\pi) - u(0)] - \int_0^{2\pi} f\left(e^{is}\right) du(s) \right| \leq 2^r H \bigvee_a^b (u).$$

$$\tag{4.21}$$

Theorem 4.3 (Dragomir 2015, [17]) *Assume that* f :

$\mathcal{C}(0,1) \rightarrow \mathbb{C}$ *satisfies the Hölder-type condition (4.6). If* $[a,b] \subseteq [0, 2\pi]$ *and the function* $u : [a,b] \rightarrow \mathbb{C}$ *is Lipschitzian with the constant* $L > 0$ *on* $[a,b]$, *then*

$$|T_{\mathcal{C}}(f, u; a, b; t)| \leq 2^r L H C_r(a, b; t) \qquad (4.22)$$

for any $t \in [a,b]$, *where*

$$C_r(a, b; t) \qquad (4.23)$$
$$:= \int_t^b \sin^r \left(\frac{b-s}{2} \right) ds + \int_a^t \sin^r \left(\frac{s-a}{2} \right) ds$$
$$\leq \frac{(b-t)^{r+1} + (t-a)^{r+1}}{(r+1) 2^r}$$

for any $t \in [a,b]$.

In particular, if $f : \mathcal{C}(0,1) \rightarrow \mathbb{C}$ *is Lipschitzian with the constant* $K > 0$, *then we have*

$$|T_{\mathcal{C}}(f, u; a, b; t)| \leq 8LK \left[\sin^2 \left(\frac{b-t}{4} \right) + \sin^2 \left(\frac{t-a}{4} \right) \right]$$
$$(4.24)$$

for any $t \in [a,b]$.

Proof. It is well known that if $p : [a,b] \rightarrow \mathbb{C}$ is a Riemann integrable function and $v : [a,b] \rightarrow \mathbb{C}$ is Lipschitzian with the constant $M > 0$, then the Riemann–Stieltjes integral $\int_a^b p(t) \, dv(t)$ exists and the following inequality holds

$$\left| \int_a^b p(t) \, dv(t) \right| \leq M \int_a^b |p(t)| \, dt. \qquad (4.25)$$

Utilizing this property and the equality (4.10) we have

$$|T_{\mathcal{C}}\left(f, u; a, b; t\right)| \tag{4.26}$$

$$\leq \left| \int_t^b \left[f\left(e^{ib}\right) - f\left(e^{is}\right)\right] du\left(s\right)\right|$$

$$+ \left| \int_a^t \left[f\left(e^{is}\right) - f\left(e^{ia}\right)\right] du\left(s\right)\right|$$

$$\leq L \left[\int_t^b \left| f\left(e^{ib}\right) - f\left(e^{is}\right)\right| ds + \int_a^t \left| f\left(e^{is}\right) - f\left(e^{ia}\right)\right| ds \right]$$

$$\leq LH \left[\int_t^b \left| e^{ib} - e^{is}\right|^r ds + \int_a^t \left| e^{is} - e^{ia}\right|^r ds \right]$$

$$= 2^r LH \left[\int_t^b \sin^r \left(\frac{b-s}{2}\right) ds + \int_a^t \sin^r \left(\frac{s-a}{2}\right) ds \right]$$

for any $t \in [a, b]$.

On making use of the elementary inequality $\sin x \leq x, x \in [0, \pi]$ we have

$$\int_t^b \sin^r \left(\frac{b-s}{2}\right) ds + \int_a^t \sin^r \left(\frac{s-a}{2}\right) ds$$

$$\leq \int_t^b \left(\frac{b-s}{2}\right)^r ds + \int_a^t \left(\frac{s-a}{2}\right)^r ds$$

$$= \frac{(b-t)^{r+1} + (t-a)^{r+1}}{(r+1) 2^r}$$

for any $t \in [a, b]$. This proves the inequality (4.23).

For $r = 1$

$$C_1 \left(a, b; t\right) := \int_t^b \sin \left(\frac{b-s}{2}\right) ds + \int_a^t \sin \left(\frac{s-a}{2}\right) ds$$

$$= 2 - 2 \cos \left(\frac{b-t}{2}\right) + 2 - 2 \cos \left(\frac{t-a}{2}\right)$$

$$= 4 \left[\sin^2 \left(\frac{b-t}{4}\right) + \sin^2 \left(\frac{t-a}{4}\right)\right]$$

for any $t \in [a, b]$.

Using (4.22) for $r = 1$ we deduce (4.24).

Remark 18 *For $a = 0$ and $b = 2\pi$ we have*

$$\sin^2\left(\frac{b-t}{4}\right) + \sin^2\left(\frac{t-a}{4}\right) = \sin^2\left(\frac{\pi}{2} - \frac{t}{4}\right) + \sin^2\left(\frac{t}{4}\right)$$
$$= \cos^2\left(\frac{t}{4}\right) + \sin^2\left(\frac{t}{4}\right) = 1$$

and by (4.24) we deduce that

$$\left| f(1)\left[u(2\pi) - u(0)\right] - \int_0^{2\pi} f\left(e^{is}\right) du(s) \right| \leq 8LK \quad (4.27)$$

for any $t \in [a, b]$.

The case of midpoint rule $t = \frac{a+b}{2}$ is as follows:

Corollary 4.4 (Dragomir 2015, [17]) *Assume that f : $\mathcal{C}(0, 1) \to \mathbb{C}$ is Lipschitzian with the constant $K > 0$ and $u : [a, b] \to \mathbb{C}$ is Lipschitzian with the constant $L > 0$ on $[a, b]$. Then we have*

$$|M_{\mathcal{C}}(f, u; a, b)| \leq 16LK \sin^2\left(\frac{b-a}{8}\right). \quad (4.28)$$

The case of monotonic nondecreasing integrators that is important for applications for unitary operators is as follows.

Theorem 4.5 (Dragomir 2015, [17]) *Assume that f : $\mathcal{C}(0, 1) \to \mathbb{C}$ satisfies the Hölder-type condition (4.6). If $[a, b] \subseteq [0, 2\pi]$ and the function $u : [a, b] \to \mathbb{R}$ is monotonic nondecreasing on $[a, b]$, then*

$$|T_{\mathcal{C}}(f, u; a, b; t)| \leq 2^r H D_r(u; a, b; t) \quad (4.29)$$

for any $t \in [a, b]$, where

$$D_r(u; a, b; t) \quad (4.30)$$
$$:= \int_t^b \sin^r\left(\frac{b-s}{2}\right) du(s) + \int_a^t \sin^r\left(\frac{s-a}{2}\right) du(s)$$
$$\leq \frac{1}{2^r}\left[\int_t^b (b-s)^r du(s) + \int_a^t (s-a)^r du(s)\right]$$

for any $t \in [a, b]$.

Proof. It is well known that if $p : [a, b] \to \mathbb{C}$ is a continuous function and $v : [a, b] \to \mathbb{R}$ is monotonic nondecreasing on $[a, b]$, then the Riemann–Stieltjes integral $\int_a^b p(t)\, dv(t)$ exists and the following inequality holds

$$\left| \int_a^b p(t)\, dv(t) \right| \leq \int_a^b |p(t)|\, dv(t). \qquad (4.31)$$

Utilizing the property (4.31), we have from (4.10) that

$$|T_{\mathcal{C}}(f, u; a, b; t)| \qquad (4.32)$$

$$\leq \left| \int_t^b \left[f\left(e^{ib}\right) - f\left(e^{is}\right) \right] du(s) \right|$$

$$+ \left| \int_a^t \left[f\left(e^{is}\right) - f\left(e^{ia}\right) \right] du(s) \right|$$

$$\leq \left[\int_t^b \left| f\left(e^{ib}\right) - f\left(e^{is}\right) \right| du(s) \right.$$

$$\left. + \int_a^t \left| f\left(e^{is}\right) - f\left(e^{ia}\right) \right| du(s) \right]$$

$$\leq H \left[\int_t^b \left| e^{ib} - e^{is} \right|^r du(s) + \int_a^t \left| e^{is} - e^{ia} \right|^r du(s) \right]$$

$$= 2^r H \left[\int_t^b \sin^r \left(\frac{b-s}{2} \right) du(s) + \int_a^t \sin^r \left(\frac{s-a}{2} \right) du(s) \right]$$

for any $t \in [a, b]$, which proves (4.29).

Moreover, by the elementary inequality $\sin x \leq x, x \in [0, \pi]$ and the monotonicity of u we also have

$$\int_t^b \sin^r \left(\frac{b-s}{2} \right) du(s) + \int_a^t \sin^r \left(\frac{s-a}{2} \right) du(s)$$

$$\leq \int_t^b \left(\frac{b-s}{2} \right)^r du(s) + \int_a^t \left(\frac{s-a}{2} \right)^r du(s)$$

which proves (4.30).

Corollary 4.6 (Dragomir 2015, [17]) *Assume that f is as* *in Theorem 4.5. If the function $u : [0, 2\pi] \to \mathbb{R}$ is monotonic nondecreasing on $[0, 2\pi]$, then*

$$\left| f(1) [u(2\pi) - u(0)] - \int_0^{2\pi} f\left(e^{is}\right) du(s) \right| \tag{4.33}$$

$$\leq 2^r H \int_0^{2\pi} \sin^r \left(\frac{s}{2}\right) du(s) = 2^{r/2} H \int_0^{2\pi} (1 - \cos s)^{r/2} du(s).$$

Proof. We have

$$
\begin{aligned}
&D_r(f, u; 0, 2\pi; t) \\
&:= \int_t^{2\pi} \sin^r \left(\frac{2\pi - s}{2}\right) du(s) + \int_0^t \sin^r \left(\frac{s}{2}\right) du(s) \\
&= \int_t^{2\pi} \sin^r \left(\pi - \frac{s}{2}\right) du(s) + \int_0^t \sin^r \left(\frac{s}{2}\right) du(s) \\
&= \int_t^{2\pi} \sin^r \left(\frac{s}{2}\right) du(s) + \int_0^t \sin^r \left(\frac{s}{2}\right) du(s) \\
&= \int_0^{2\pi} \sin^r \left(\frac{s}{2}\right) du(s)
\end{aligned}
$$

for any $t \in [0, 2\pi]$.

Since for $s \in [0, 2\pi]$ we have

$$\sin\left(\frac{s}{2}\right) = \left(\frac{1 - \cos s}{2}\right)^{1/2}$$

then the last part of (4.33) is obtained.

4.3 APPLICATIONS FOR FUNCTIONS OF UNITARY OPERATORS

We have the following vector inequality for functions of unitary operators.

Theorem 4.7 (Dragomir 2015, [17]) *Assume that $f : \mathcal{C}(0, 1) \to \mathbb{C}$ satisfies the Hölder-type condition (4.6). If the operator $U : H \to H$ on the Hilbert space H is unitary and*

$\{E_\lambda\}_{\lambda \in [0,2\pi]}$ *is its spectral family, then*

$$|f(1)\langle x,y \rangle - \langle f(U)x,y \rangle| \le 2^r H \bigvee_0^{2\pi} \left(\langle E_{(\cdot)}x,y \rangle \right)$$

$$\le 2^r H \|x\| \|y\| \qquad (4.34)$$

for any $x,y \in H.$

Proof. For given $x,y \in H$, define the function $u(\lambda) := \langle E_\lambda x,y \rangle, \lambda \in [0,2\pi]$. We know that u is of bounded variation and

$$\bigvee_0^{2\pi}(u) =: \bigvee_0^{2\pi}\left(\langle E_{(\cdot)}x,y \rangle\right) \le \|x\|\|y\|. \qquad (4.35)$$

On making use of the spectral representation theorem and the inequality (4.35) we deduce the desired result (4.34).

Theorem 4.8 (Dragomir 2015, [17]) *With the assumptions of Theorem 4.7 we have*

$$\left| f(1)\|x\|^2 - \langle f(U)x,x \rangle \right| \le 2^{r/2}H\left\langle [1_H - \mathrm{Re}(U)]^{r/2}x,x \right\rangle, \qquad (4.36)$$

for any $x \in H$, *where*

$$\mathrm{Re}(U) := \frac{U + U^*}{2}.$$

Proof. Utilizing the inequality (4.33), we have

$$\left| f(1)[\langle E_{2\pi}x,x \rangle - \langle E_0 x,x \rangle] - \int_0^{2\pi} f\left(e^{is}\right)d\langle E_s x,x \rangle \right| \qquad (4.37)$$

$$\le 2^{r/2}H\int_0^{2\pi}(1-\cos s)^{r/2}\,d\langle E_s x,x \rangle$$

for any $x \in H.$

Since

$$\int_0^{2\pi}(1-\cos s)^{r/2}\,d\langle E_s x,x \rangle = \int_0^{2\pi}\left(1 - \mathrm{Re}\left(e^{is}\right)\right)^{r/2}d\langle E_s x,x \rangle$$

$$= \left\langle (1_H - \mathrm{Re}(U))^{r/2}x,x \right\rangle$$

for any $x \in H$, then by (4.37) we get the desired result (4.36).

Example 4.1 *In order to provide some simple examples for the inequalities above, we choose two complex functions as follows.*

a) *Consider the power function* $f : \mathbb{C} \backslash \{0\} \to \mathbb{C}$, $f(z) = z^m$ *where m is a nonzero integer. Then, obviously, for any z, w belonging to the unit circle $\mathcal{C}(0,1)$ we have the inequality*

$$|f(z) - f(w)| \leq |m| \, |z - w|$$

which shows that f is Lipschitzian with the constant $L = |m|$ on the circle $\mathcal{C}(0,1)$.

Then from (4.36), we get for any unitary operator U that

$$\left| \|x\|^2 - \langle U^m x, x \rangle \right| \leq 2^{1/2} |m| \left\langle (1_H - \mathrm{Re}\,(U))^{1/2} x, x \right\rangle, \tag{4.38}$$

for any $x \in H$.

For $m = 1$ we also get from (4.34) that

$$|\langle x, y \rangle - \langle Ux, y \rangle| \leq 2 \bigvee_0^{2\pi} \left(\left\langle E_{(\cdot)} x, y \right\rangle \right) \leq 2 \|x\| \, \|y\| \tag{4.39}$$

for any $x, y \in H$.

b) *For $a \neq \pm 1, 0$ consider the function $f : \mathcal{C}(0,1) \to \mathbb{C}$, $f_a(z) = \frac{1}{1-az}$. Observe that*

$$|f_a(z) - f_a(w)| = \frac{|a| \, |z - w|}{|1 - az| \, |1 - aw|} \tag{4.40}$$

for any $z, w \in \mathcal{C}(0,1)$.

If $z = e^{it}$ with $t \in [0, 2\pi]$, then we have

$$
\begin{aligned}
|1 - az|^2 &= 1 - 2a \, \mathrm{Re}\,(\bar{z}) + a^2 |z|^2 = 1 - 2a \cos t + a^2 \\
&\geq 1 - 2|a| + a^2 = (1 - |a|)^2
\end{aligned}
$$

therefore

$$\frac{1}{|1-az|} \le \frac{1}{|1-|a||} \quad and \quad \frac{1}{|1-aw|} \le \frac{1}{|1-|a||} \quad (4.41)$$

for any $z, w \in C(0,1)$.

Utilizing (4.40) and (4.41) we deduce

$$|f_a(z) - f_a(w)| \le \frac{|a|}{(1-|a|)^2} |z - w| \quad (4.42)$$

for any $z, w \in C(0,1)$, *showing that the function* f_a *is Lipschitzian with the constant* $L_a = \frac{|a|}{(1-|a|)^2}$ *on the circle* $C(0,1)$.

Now, if we employ the inequality (4.34), then we can state the inequality

$$\left| (1-a)^{-1} \langle x, y \rangle - \left\langle (1_H - aU)^{-1} x, y \right\rangle \right| \quad (4.43)$$

$$\le \frac{2|a|}{(1-|a|)^2} \bigvee_0^{2\pi} \left(\left\langle E_{(\cdot)} x, y \right\rangle \right)$$

$$\le \frac{2|a|}{(1-|a|)^2} \|x\| \|y\|$$

for any unitary operator U *and for any* $x, y \in H$.

On making use of the inequality (4.36) we also have

$$\left| (1-a)^{-1} \|x\|^2 - \left\langle (1_H - aU)^{-1} x, x \right\rangle \right| \quad (4.44)$$

$$\le \frac{2^{1/2} |a|}{(1-|a|)^2} \left\langle [1_H - \mathrm{Re}(U)]^{1/2} x, x \right\rangle,$$

for any $x \in H$.

4.4   A QUADRATURE RULE

We consider the following partition of the interval $[a, b]$

$$\Delta_n : a = x_0 < x_1 < ... < x_{n-1} < x_n = b$$

and the intermediate points $\xi_k \in [x_k, x_{k+1}]$ where $0 \leq k \leq n-1$. Define $h_k := x_{k+1} - x_k$, $0 \leq k \leq n-1$ and $\nu(\Delta_n) = \max\{h_k : 0 \leq k \leq n-1\}$, the norm of the partition Δ_n.

For the continuous function $f : \mathcal{C}(0,1) \to \mathbb{C}$ and the function $u : [a, b] \subseteq [0, 2\pi] \to \mathbb{C}$ of bounded variation on $[a, b]$, define the *generalized trapezoid quadrature rule*

$$T_n(f, u, \Delta_n, \xi) \tag{4.45}$$

$$:= \sum_{k=0}^{n-1} \left\{ f\left(e^{ix_{k+1}}\right) [u(x_{k+1}) - u(\xi_k)] \right.$$

$$\left. + f\left(e^{ix_k}\right) [u(\xi_k) - u(x_k)] \right\}$$

and the remainder $R_n(f, u, \Delta_n, \xi)$ in approximating the Riemann–Stieltjes integral $\int_a^b f\left(e^{it}\right) du(t)$ by $T_n(f, u, \Delta_n, \xi)$. Then we have

$$\int_a^b f\left(e^{it}\right) du(t) = T_n(f, u, \Delta_n, \xi) + R_n(f, u, \Delta_n, \xi). \tag{4.46}$$

The following result provides *a priori* bounds for $R_n(f, u, \Delta_n, \xi)$ in several instances of f and u as above.

Proposition 4.9 (Dragomir 2015, [17]) *Assume that $f : \mathcal{C}(0,1) \to \mathbb{C}$ satisfies the following Hölder-type condition*

$$|f(z) - f(w)| \leq H|z - w|^r$$

for any $w, z \in \mathcal{C}(0,1)$, where $H > 0$ and $r \in (0, 1]$ are given.

If $[a, b] \subseteq [0, 2\pi]$ and the function $u : [a, b] \to \mathbb{C}$ is of bounded variation on $[a, b]$, then for any partition $\Delta_n : a = x_0 < x_1 < ... < x_{n-1} < x_n = b$ with the norm $\nu(\Delta_n) \leq \pi$ we have the error bound

$$|R_n\left(f,u,\Delta_n,\xi\right)| \qquad\qquad (4.47)$$

$$\leq 2^r H \sum_{k=0}^{n-1} \sin^r \left(\frac{1}{2}\left[\frac{x_{k+1}-x_k}{2}+\left|\xi_k-\frac{x_{k+1}+x_k}{2}\right|\right]\right) \overset{x_{k+1}}{\underset{x_k}{\bigvee}}(u)$$

$$\leq 2^r H \sum_{k=0}^{n-1} \sin^r \left(\frac{x_{k+1}-x_k}{2}\right) \overset{x_{k+1}}{\underset{x_k}{\bigvee}}(u)$$

$$\leq 2^r \sum_{k=0}^{n-1} \sin^r \left(\frac{x_{k+1}-x_k}{2}\right) \overset{x_{k+1}}{\underset{x_k}{\bigvee}}(u)$$

$$\leq 2^r H \sin^r \left(\frac{\nu\left(\Delta_n\right)}{2}\right) \overset{b}{\underset{a}{\bigvee}}(u) \leq \nu^r\left(\Delta_n\right) H \overset{b}{\underset{a}{\bigvee}}(u)$$

for any intermediate points $\xi_k \in [x_k, x_{k+1}]$ *where* $0 \leq k \leq n-1.$

Proof. Since $\nu\left(\Delta_n\right) \leq \pi$, then on writing inequality (4.20) on each interval $[x_k, x_{k+1}]$ and for any intermediate points

$\xi_k \in [x_k, x_{k+1}]$ where $0 \le k \le n-1$, we have

$$
\left| \int_{x_k}^{x_{k+1}} f\left(e^{it}\right) du\left(t\right) \right. \tag{4.48}
$$

$$
\left. - f\left(e^{ix_{k+1}}\right) \left[u\left(x_{k+1}\right) - u\left(\xi_k\right)\right] - f\left(e^{ix_k}\right) \left[u\left(\xi_k\right) - u\left(x_k\right)\right] \right|
$$

$$
\le 2^r H \left[\sin^r \left(\frac{\xi_k - x_k}{2} \right) \bigvee_{x_k}^{\xi_k} (u) + \sin^r \left(\frac{x_{k+1} - \xi_k}{2} \right) \bigvee_{\xi_k}^{x_{k+1}} (u) \right]
$$

$$
\le 2^r H \max \left\{ \sin^r \left(\frac{\xi_k - x_k}{2} \right), \sin^r \left(\frac{x_{k+1} - \xi_k}{2} \right) \right\} \bigvee_{x_k}^{x_{k+1}} (u)
$$

$$
= 2^r H \sin^r \left(\max \left\{ \left(\frac{\xi_k - x_k}{2} \right), \left(\frac{x_{k+1} - \xi_k}{2} \right) \right\} \right) \bigvee_{x_k}^{x_{k+1}} (u)
$$

$$
\le 2^r H \sin^r \left(\frac{1}{2} \left[\frac{x_{k+1} - x_k}{2} + \left| \xi_k - \frac{x_{k+1} + x_k}{2} \right| \right] \right) \bigvee_{x_k}^{x_{k+1}} (u)
$$

$$
\le 2^r H \sin^r \left(\frac{x_{k+1} - x_k}{2} \right) \bigvee_{x_k}^{x_{k+1}} (u) \le 2^r H \sin^r \left(\frac{x_{k+1} - x_k}{2} \right)
$$

$$
\times \bigvee_{x_k}^{x_{k+1}} (u)
$$

$$
\le 2^r H \sin^r \left(\frac{\nu\left(\Delta_n\right)}{2} \right) \bigvee_{x_k}^{x_{k+1}} (u) \le \nu^r\left(\Delta_n\right) H \bigvee_{x_k}^{x_{k+1}} (u).
$$

Summing over k from 0 to $n-1$ in (4.48) and utilizing the generalized triangle inequality, we deduce (4.47).

For the continuous function $f : \mathcal{C}\left(0, 1\right) \to \mathbb{C}$ and the function $u : [a, b] \subseteq [0, 2\pi] \to \mathbb{C}$ of bounded variation on $[a, b]$, define the *trapezoid midpoint quadrature rule*

$$
M_n\left(f, u, \Delta_n\right) := \sum_{k=0}^{n-1} f\left(e^{ix_{k+1}}\right) \left[u\left(x_{k+1}\right) - u\left(\frac{x_k + x_{k+1}}{2} \right)\right]
$$

$$
\tag{4.49}
$$

$$
+ \sum_{k=0}^{n-1} f\left(e^{ix_k}\right) \left[u\left(\frac{x_k + x_{k+1}}{2} \right) - u\left(x_k\right)\right]
$$

and the remainder $T_n (f, u, \Delta_n)$ in approximating the Riemann–Stieltjes integral $\int_a^b f\left(e^{it}\right) du\left(t\right)$ by $M_n\left(f, u, \Delta_n\right)$. Then we have

$$\int_a^b f\left(e^{it}\right) du\left(t\right) = M_n\left(f, u, \Delta_n\right) + T_n\left(f, u, \Delta_n\right). \quad (4.50)$$

Proposition 4.10 (Dragomir 2015, [17]) *Assume that f and Δ_n are as in Proposition 4.9, then we have the error bound*

$$|T_n\left(f, u, \Delta_n\right)| \leq 2^r H \sum_{k=0}^{n-1} \sin^r\left(\frac{x_{k+1} - x_k}{4}\right) \bigvee_{x_k}^{x_{k+1}}\left(u\right) \quad (4.51)$$

$$\leq 2^r H \sin^r\left(\frac{\nu\left(\Delta_n\right)}{4}\right) \bigvee_a^b\left(u\right) \leq \frac{1}{2^r}\nu^r\left(\Delta_n\right) H \bigvee_a^b\left(u\right).$$

We consider the following partition of the interval $[0, 2\pi]$

$$\Gamma_n : 0 = \lambda_0 < \lambda_1 < \ldots < \lambda_{n-1} < \lambda_n - 2\pi$$

and the intermediate points $\xi_k \in [\lambda_k, \lambda_{k+1}]$ where $0 \leq k \leq n-1$. Define $h_k := \lambda_{k+1} - \lambda_k$, $0 \leq k \leq n-1$ and $\nu\left(\Gamma_n\right) = \max\{h_k : 0 \leq k \leq n-1\}$, the norm of the partition Γ_n.

If U is a unitary operator on the Hilbert space H and $\{E_\lambda\}_{\lambda \in [0, 2\pi]}$, the spectral family of U, then we can introduce the following sums

$$T_n\left(f, \Gamma_n, \xi; x, y\right) \quad (4.52)$$

$$:= \sum_{k=0}^{n-1} \left\{ f\left(e^{i\lambda_{k+1}}\right) \langle\left(E_{\lambda_{k+1}} - E_{\xi_k}\right) x, y\rangle \right.$$

$$\left. + f\left(e^{i\lambda_k}\right) \langle\left(E_{\xi_k} - E_{\lambda_k}\right) x, y\rangle \right\}$$

for $x, y \in H$.

For a function $f : \mathcal{C}\left(0, 1\right) \to \mathbb{C}$ that satisfies the Hölder-type condition (4.6), we can approximate the function f of unitary operator U as follows

$$\langle f\left(U\right) x, y\rangle = T_n\left(f, \Gamma_n, \xi; x, y\right) + R_n\left(f, \Gamma_n, \xi; x, y\right) \quad (4.53)$$

for $x, y \in H$, where the remainder satisfies the bounds

$$|R_n \left(f, \Gamma_n, \xi; x, y\right)| \tag{4.54}$$

$$\leq 2^r H \sum_{k=0}^{n-1} \sin^r \left(\frac{1}{2}\left[\frac{\lambda_{k+1} - \lambda_k}{2} + \left|\xi_k - \frac{\lambda_{k+1} + \lambda_k}{2}\right|\right]\right)$$

$$\times \bigvee_{\lambda_k}^{\lambda_{k+1}} \left(\left\langle E_{(\cdot)}x, y\right\rangle\right)$$

$$\leq 2^r H \sum_{k=0}^{n-1} \sin^r \left(\frac{\lambda_{k+1} - \lambda_k}{2}\right) \bigvee_{\lambda_k}^{\lambda_{k+1}} \left(\left\langle E_{(\cdot)}x, y\right\rangle\right)$$

$$\leq 2^r \sum_{k=0}^{n-1} \sin^r \left(\frac{\lambda_{k+1} - \lambda_k}{2}\right) \bigvee_{\lambda_k}^{\lambda_{k+1}} \left(\left\langle E_{(\cdot)}x, y\right\rangle\right)$$

$$\leq 2^r H \sin^r \left(\frac{\nu\left(\Gamma_n\right)}{2}\right) \bigvee_{0}^{2\pi} \left(\left\langle E_{(\cdot)}x, y\right\rangle\right) \leq \nu^r \left(\Gamma_n\right)$$

$$\times H \bigvee_{0}^{2\pi} \left(\left\langle E_{(\cdot)}x, y\right\rangle\right)$$

for any $x, y \in H$.

The interested reader may apply the above results for various Lipschitzian functions $f : \mathcal{C}\left(0, 1\right) \to \mathbb{C}$. However, the details are not presented here.

CHAPTER **5**

Quasi-Grüss-Type Inequalities

$\mathbf{IN}$ THIS CHAPTER we present some Riemann–Stieltjes integral inequalities of quasi-Grüss type for continuous complex-valued integrands and various classes of bounded variation integrators. Several applications for functions of unitary operators in Hilbert spaces are provided as well.

5.1 INTRODUCTION

The concept of *Riemann–Stieltjes integral* $\int_a^b f(t)\,du(t)$, where f is called *the integrand* and u is called *the integrator*, plays an important role in Mathematics, for instance in the definition of a complex integral, the representation of bounded linear functionals on the Banach space of all continuous functions on an interval $[a, b]$, in the spectral representation of self-adjoint operators on complex Hilbert spaces and other classes of operators such as the unitary operators etc.

One can approximate the Riemann–Stieltjes integral $\int_a^b f(t)\,du(t)$ with the following simpler quantity:

$$\frac{1}{b-a}\left[u(b)-u(a)\right]\cdot\int_a^b f(t)\,dt \qquad ([25], [26]). \qquad (5.1)$$

In order to provide *a priori* sharp bounds for the *approximation error*, consider the functionals:

$$D\left(f, u; a, b\right) := \int_a^b f\left(t\right) du\left(t\right) - \frac{1}{b-a}\left[u\left(b\right) - u\left(a\right)\right] \cdot \int_a^b f\left(t\right) dt.$$

If the integrand f is *Riemann integrable* on $[a, b]$ and the integrator $u : [a, b] \to \mathbb{R}$ is $L-Lipschitzian$, i.e.,

$$\left|u\left(t\right) - u\left(s\right)\right| \leq L\left|t - s\right| \qquad \text{for each } t, s \in [a, b], \qquad (5.2)$$

then the Riemann–Stieltjes integral $\int_a^b f\left(t\right) du\left(t\right)$ exists and, as pointed out in [25], the following quasi-Grüss-type inequality holds

$$\left|D\left(f, u; a, b\right)\right| \leq L \int_a^b \left|f\left(t\right) - \int_a^b \frac{1}{b-a} f\left(s\right) ds\right| dt. \qquad (5.3)$$

The inequality (5.3) is sharp in the sense that the multiplicative constant $C = 1$ in front of L cannot be replaced by a smaller quantity. Moreover, if there exist the constants $m, M \in \mathbb{R}$ such that $m \leq f\left(t\right) \leq M$ for a.e. $t \in [a, b]$, then [25]

$$\left|D\left(f, u; a, b\right)\right| \leq \frac{1}{2}L\left(M - m\right)\left(b - a\right). \qquad (5.4)$$

The constant $\frac{1}{2}$ is the best possible in (5.4).

We call this type of inequality a *quasi-Grüss type* since for integrators of integral form $u\left(t\right) := \frac{1}{b-a}\int_a^t g\left(s\right) ds$, the left-hand side becomes

$$\left|\frac{1}{b-a}\int_a^b f\left(t\right) g\left(t\right) dt - \frac{1}{b-a}\int_a^b f\left(t\right) dt \cdot \frac{1}{b-a}\int_a^b g\left(s\right) ds\right|,$$

which is related to the well-known Grüss inequality.

A different approach in the case of integrands of bounded variation were considered by the same authors in 2001, [26], where they showed that

$$\left|D\left(f, u; a, b\right)\right| \leq \max_{t \in [a,b]}\left|f\left(t\right) - \frac{1}{b-a}\int_a^b f\left(s\right) ds\right| \bigvee_a^b\left(u\right), \qquad (5.5)$$

provided that f is continuous and u is of bounded variation. Here $\bigvee_a^b (u)$ denotes the total variation of u on $[a, b]$. The inequality (5.5) is sharp.

If we assume that f is K−Lipschitzian, then [26]

$$|D(f, u; a, b)| \leq \frac{1}{2} K (b - a) \bigvee_a^b (u),$$ (5.6)

with $\frac{1}{2}$ the best possible constant in (5.6).

For various bounds on the error functional $D(f, u; a, b)$ where f and u belong to different classes of function for which the Stieltjes integral exists, see [16], [15], [14], and [6] and the references therein.

For continuous functions $f : \mathcal{C}(0, 1) \to \mathbb{C}$, where $\mathcal{C}(0, 1)$ is the unit circle from $\mathbb{C}$ centered in 0 and $u : [a, b] \subseteq [0, 2\pi] \to \mathbb{C}$ is a function of bounded variation on $[a, b]$, we can define the following *functional of quasi-Grüss* type as well:

$$D_{\mathcal{C}}(f; u, a, b)$$ (5.7)

$$:= \int_a^b f\left(e^{it}\right) du(t) - \frac{1}{b - a} [u(b) - u(a)] \cdot \int_a^b f\left(e^{it}\right) dt.$$

In this paper we establish some bounds for the magnitude of $S_{\mathcal{C}}(f; u, a, b)$ when the integrand $f : \mathcal{C}(0, 1) \to \mathbb{C}$ satisfies some *Hölder-type conditions* on the circle $\mathcal{C}(0, 1)$ while the integrator u is of bonded variation.

5.2 INEQUALITIES FOR RIEMANN–STIELTJES INTE-GRAL

We say that the complex function $f : \mathcal{C}(0, 1) \to \mathbb{C}$ satisfies an *H-r-Hölder-type condition* on the circle $\mathcal{C}(0, 1)$, where $H > 0$ and $r \in (0, 1]$ are given, if

$$|f(z) - f(w)| \leq H |z - w|^r$$ (5.8)

for any $w, z \in \mathcal{C}(0, 1)$.

If $r = 1$ and $L = H$ then we call it of *L-Lipschitz type*.

Consider the power function $f : \mathbb{C}\backslash\{0\} \to \mathbb{C}$, $f(z) = z^m$ where m is a nonzero integer. Then, obviously, for any z, w belonging to the unit circle $\mathcal{C}(0, 1)$ we have the inequality

$$|f(z) - f(w)| \leq |m| \, |z - w|$$

which shows that f is Lipschitzian with the constant $L = |m|$ on the circle $\mathcal{C}(0, 1)$.

For $a \neq \pm 1, 0$ real numbers, consider the function $f : \mathcal{C}(0, 1) \to \mathbb{C}$, $f_a(z) = \frac{1}{1-az}$. Observe that

$$|f_a(z) - f_a(w)| = \frac{|a| \, |z - w|}{|1 - az| \, |1 - aw|} \tag{5.9}$$

for any $z, w \in \mathcal{C}(0, 1)$.

If $z = e^{it}$ with $t \in [0, 2\pi]$, then we have

$$
\begin{aligned}
|1 - az|^2 &= 1 - 2a \operatorname{Re}(\bar{z}) + a^2 |z|^2 = 1 - 2a \cos t + a^2 \\
&\geq 1 - 2|a| + a^2 = (1 - |a|)^2
\end{aligned}
$$

therefore

$$\frac{1}{|1 - az|} \leq \frac{1}{|1 - |a||} \quad \text{and} \quad \frac{1}{|1 - aw|} \leq \frac{1}{|1 - |a||} \tag{5.10}$$

for any $z, w \in \mathcal{C}(0, 1)$.

Utilizing (5.9) and (5.10) we deduce

$$|f_a(z) - f_a(w)| \leq \frac{|a|}{(1 - |a|)^2} |z - w| \tag{5.11}$$

for any $z, w \in \mathcal{C}(0, 1)$, showing that the function f_a is Lipschitzian with the constant $L_a = \frac{|a|}{(1-|a|)^2}$ on the circle $\mathcal{C}(0, 1)$.

Theorem 5.1 (Dragomir 2016, [20]) *Let $f : \mathcal{C}(0, 1) \to \mathbb{C}$ satisfy an H-r-Hölder-type condition on the circle $\mathcal{C}(0, 1)$, where $H > 0$ and $r \in (0, 1]$ are given. If $u : [a, b] \subseteq [0, 2\pi] \to \mathbb{C}$ is a function of bounded variation on $[a, b]$, then*

$$|D_{\mathcal{C}}(f; u, a, b)| \leq \frac{2^r H}{b - a} \max_{t \in [a,b]} B_r(a, b; t) \bigvee_a^b (u) \tag{5.12}$$

$$\leq \frac{H}{r + 1} (b - a)^r \bigvee_a^b (u)$$

where

$$B_r(a, b; t) := \int_a^t \sin^r\left(\frac{t-s}{2}\right) ds + \int_t^b \sin^r\left(\frac{s-t}{2}\right) ds \tag{5.13}$$

$$\leq \frac{1}{2^r} \frac{(t-a)^{r+1} + (b-t)^{r+1}}{r+1}$$

for any $t \in [a, b]$.

In particular, if f is Lipschitzian with the constant $L > 0$, and $[a, b] \subset [0, 2\pi]$ with $b - a \neq 2\pi$, then we have the simpler inequality

$$|D_{\mathcal{C}}(f; u, a, b)| \leq \frac{8L}{b-a} \sin^2\left(\frac{b-a}{4}\right) \bigvee_a^b(u) \leq \frac{1}{2} L(b-a) \bigvee_a^b(u). \tag{5.14}$$

If $a = 0$ and $b = 2\pi$ and f is Lipschitzian with the constant $L > 0$, then

$$|D_{\mathcal{C}}(f; u, 0, 2\pi)| \leq \frac{4L}{\pi} \bigvee_0^{2\pi}(u). \tag{5.15}$$

Proof. We have

$$D_{\mathcal{C}}(f; u, a, b) = \int_a^b \left(f\left(e^{it}\right) - \frac{1}{b-a} \int_a^b f\left(e^{is}\right) ds \right) du(t) \tag{5.16}$$

$$= \frac{1}{b-a} \int_a^b \left(\int_a^b \left[f\left(e^{it}\right) - f\left(e^{is}\right) \right] ds \right) du(t).$$

It is known that if $p : [c, d] \to \mathbb{C}$ is a continuous function and $v : [c, d] \to \mathbb{C}$ is of bounded variation, then the Riemann–Stieltjes integral $\int_c^d p(t) \, dv(t)$ exists and the following inequality holds

$$\left| \int_c^d p(t) \, dv(t) \right| \leq \max_{t \in [c,d]} |p(t)| \bigvee_c^d(v). \tag{5.17}$$

Utilizing this property and (5.16) we have

$$
|D_{\mathcal{C}}(f; u, a, b)| = \frac{1}{b-a} \left| \int_a^b \left(\int_a^b \left[f\left(e^{it}\right) - f\left(e^{is}\right) \right] ds \right) du(t) \right|
$$

$$
\tag{5.18}
$$

$$
\leq \frac{1}{b-a} \max_{x \in [a,b]} \left| \int_a^b \left[f\left(e^{it}\right) - f\left(e^{is}\right) \right] ds \right| \bigvee_a^b (u).
$$

Utilizing the properties of the Riemann integral and the fact that f is of H-r-Hölder-type on the circle $\mathcal{C}(0,1)$ we have

$$
\left| \int_a^b \left[f\left(e^{it}\right) - f\left(e^{is}\right) \right] ds \right| \leq \int_a^b \left| f\left(e^{it}\right) - f\left(e^{is}\right) \right| ds \tag{5.19}
$$

$$
\leq H \int_a^b \left| e^{is} - e^{it} \right|^r ds.
$$

Since

$$
\left| e^{is} - e^{it} \right|^2 = \left| e^{is} \right|^2 - 2\operatorname{Re}\left(e^{i(s-t)} \right) + \left| e^{it} \right|^2
$$

$$
= 2 - 2\cos(s-t) = 4\sin^2\left(\frac{s-t}{2} \right)
$$

for any $t, s \in \mathbb{R}$, then

$$
\left| e^{is} - e^{it} \right|^r = 2^r \left| \sin\left(\frac{s-t}{2} \right) \right|^r \tag{5.20}
$$

for any $t, s \in \mathbb{R}$.

Therefore

$$
\int_a^b \left| e^{is} - e^{it} \right|^r ds = 2^r \int_a^b \left| \sin\left(\frac{s-t}{2} \right) \right|^r ds \tag{5.21}
$$

$$
= 2^r \left[\int_a^t \sin^r\left(\frac{t-s}{2} \right) ds + \int_t^b \sin^r\left(\frac{s-t}{2} \right) ds \right]
$$

for any $t \in [a, b]$.

On making use of (5.19) and (5.21) we have

$$\max_{x\in[a,b]} \left| \int_a^b \left[f\left(e^{it}\right) - f\left(e^{is}\right) \right] ds \right| \le 2^r H \max_{t\in[a,b]} B_r\left(a,b;t\right)$$

and the first inequality in (5.12) is proved.

Utilizing the elementary inequality $|\sin(x)| \le |x|$, $x \in \mathbb{R}$ we have

$$B_r\left(a,b;t\right) \le \int_a^t \left(\frac{t-s}{2}\right)^r ds + \int_t^b \left(\frac{s-t}{2}\right)^r ds \qquad (5.22)$$

$$= \frac{1}{2^r} \frac{(t-a)^{r+1} + (b-t)^{r+1}}{r+1}$$

for any $t \in [a,b]$, and the inequality (5.13) is proved.

If we consider the auxiliary function $\varphi : [a,b] \to \mathbb{R}$,

$$\varphi\left(t\right) = (t-a)^{r+1} + (b-t)^{r+1}, r \in (0,1]$$

then

$$\varphi'\left(t\right) = (r+1)\left[(t-a)^r - (b-t)^r\right]$$

and

$$\varphi''\left(t\right) = (r+1)\,r\left[(t-a)^{r-1} + (b-t)^{r-1}\right].$$

We have $\varphi'\left(t\right) = 0$ iff $t = \frac{a+b}{2}$, $\varphi'\left(t\right) < 0$ for $t \in \left(a, \frac{a+b}{2}\right)$ and $\varphi'\left(t\right) > 0$ for $t \in \left(\frac{a+b}{2}, b\right)$. We also have that $\varphi''\left(t\right) > 0$ for any $t \in (a,b)$ showing that φ is strictly decreasing on $\left(a, \frac{a+b}{2}\right)$ and strictly increasing on $\left(\frac{a+b}{2}, b\right)$. We also have that

$$\min_{t\in[a,b]} \varphi\left(t\right) = \varphi\left(\frac{a+b}{2}\right) = \frac{(b-a)^{r+1}}{2^r}$$

and

$$\max_{t\in[a,b]} \varphi\left(t\right) = \varphi\left(a\right) = \varphi\left(b\right) = (b-a)^{r+1}.$$

Taking the maximum over $t \in [a,b]$ in (5.22) we deduce the second inequality in (5.12).

For $r = 1$ we have

$$B\left(a, b; t\right) := \int_a^t \sin\left(\frac{t-s}{2}\right) ds + \int_t^b \sin\left(\frac{s-t}{2}\right) ds$$

$$= 2 - 2\cos\left(\frac{t-a}{2}\right) - 2\cos\left(\frac{b-t}{2}\right) + 2$$

$$= 2\left[1 - \cos\left(\frac{t-a}{2}\right) + 1 - \cos\left(\frac{b-t}{2}\right)\right]$$

$$= 2\left[2\sin^2\left(\frac{t-a}{4}\right) + 2\sin^2\left(\frac{b-t}{4}\right)\right]$$

$$= 4\left[\sin^2\left(\frac{t-a}{4}\right) + \sin^2\left(\frac{b-t}{4}\right)\right]$$

for any $t \in [a, b]$.

Now, if we take the derivative in the first equality, we have

$$B'\left(a, b; t\right) = \sin\left(\frac{t-a}{2}\right) - \sin\left(\frac{b-t}{2}\right)$$

$$= 2\sin\left(\frac{t - \frac{a+b}{2}}{2}\right)\cos\left(\frac{b-a}{4}\right),$$

for $[a, b] \subset [0, 2\pi]$ and $b - a \neq 2\pi$.

We observe that $B'\left(a, b; t\right) = 0$ iff $t = \frac{a+b}{2}$, $B'\left(a, b; t\right) < 0$ for $t \in \left(a, \frac{a+b}{2}\right)$ and $B'\left(a, b; t\right) > 0$ for $t \in \left(\frac{a+b}{2}, b\right)$. The second derivative is given by

$$B''\left(a, b; t\right) = \cos\left(\frac{t - \frac{a+b}{2}}{2}\right)\cos\left(\frac{b-a}{4}\right)$$

and we observe that $B''\left(a, b; t\right) > 0$ for $t \in (a, b)$.

Therefore the function $B\left(a, b; \cdot\right)$ is strictly decreasing on $\left(a, \frac{a+b}{2}\right)$ and strictly increasing on $\left(\frac{a+b}{2}, b\right)$. It is also a strictly convex function on (a, b). We have

$$\min_{t \in [a,b]} B\left(a, b; t\right) = B\left(a, b; \frac{a+b}{2}\right) = 8\sin^2\left(\frac{b-a}{8}\right)$$

and

$$\max_{t \in [a,b]} B\left(a, b; t\right) = B\left(a, b; a\right) = B\left(a, b; b\right) = 4\sin^2\left(\frac{b-a}{4}\right).$$

This proves the bound (5.14).

If $a = 0$ and $b = 2\pi$, then

$$
\begin{aligned}
B(0, 2\pi; t) &= 4\left[\sin^2\left(\frac{t}{4}\right) + \sin^2\left(\frac{2\pi - t}{4}\right)\right] \\
&= 4
\end{aligned}
$$

and by (5.12) we get (5.15).

The proof is complete.

The following result also holds:

Theorem 5.2 (Dragomir 2016, [20]) *Let $f : \mathcal{C}(0,1) \to \mathbb{C}$ satisfy an H-r-Hölder-type condition on the circle $\mathcal{C}(0,1)$, where $H > 0$ and $r \in (0,1]$ are given. If $u : [a,b] \subseteq [0,2\pi] \to \mathbb{C}$ is a function of Lipschitz type with the constant $K > 0$ on $[a,b]$, then*

$$
|D_{\mathcal{C}}(f; u, a, b)| \leq \frac{2^r H K}{b-a} C_r(a,b) \leq \frac{2HK(b-a)^{r+1}}{(r+1)(r+2)} \quad (5.23)
$$

where

$$
C_r(a,b) := \int_a^b \int_a^t \sin^r\left(\frac{t-s}{2}\right) ds\,dt + \int_a^b \int_t^b \sin^r\left(\frac{s-t}{2}\right) ds\,dt
$$

$$
(5.24)
$$

$$
\leq \frac{(b-a)^{r+2}}{2^{r-1}(r+1)(r+2)}.
$$

In particular, if f is Lipschitzian with the constant $L > 0$, then we have the simpler inequality

$$
|D_{\mathcal{C}}(f; u, a, b)| \leq \frac{16LK}{b-a}\left[\frac{b-a}{2} - \sin\left(\frac{b-a}{2}\right)\right] \quad (5.25)
$$

$$
\leq \frac{LK(b-a)^2}{3}.
$$

Proof. It is well known that if $p : [c,d] \to \mathbb{C}$ is a Riemann integrable function and $v : [c,d] \to \mathbb{C}$ is Lipschitzian with the constant $M > 0$, then the Riemann–Stieltjes integral $\int_c^d p\,(t)\,dv\,(t)$ exists and the following inequality holds

$$\left| \int_c^d p\,(t)\,dv\,(t) \right| \le M \int_c^d |p\,(t)|\,dt. \tag{5.26}$$

Utilizing the equality (5.16) and this property we have

$$|D_C\,(f; u, a, b)| = \frac{1}{b-a} \left| \int_a^b \left(\int_a^b \left[f\left(e^{it}\right) - f\left(e^{is}\right) \right] ds \right) du\,(t) \right| \tag{5.27}$$

$$\le \frac{K}{b-a} \int_a^b \left| \left(\int_a^b \left[f\left(e^{it}\right) - f\left(e^{is}\right) \right] ds \right) \right| dt.$$

From (5.19) and (5.21) we have

$$\left| \int_a^b \left[f\left(e^{it}\right) - f\left(e^{is}\right) \right] ds \right| \tag{5.28}$$

$$\le \int_a^b \left| f\left(e^{it}\right) - f\left(e^{is}\right) \right| ds$$

$$\le H \int_a^b \left| e^{is} - e^{it} \right|^r ds$$

$$= 2^r H \left[\int_a^t \sin^r \left(\frac{t-s}{2} \right) ds + \int_t^b \sin^r \left(\frac{s-t}{2} \right) ds \right]$$

and by (5.27) we deduce the first part of (5.23).

Since, by (5.22), we have

$$\int_a^t \left(\frac{t-s}{2} \right)^r ds + \int_t^b \left(\frac{s-t}{2} \right)^r ds$$

$$= \frac{1}{2^r} \frac{(t-a)^{r+1} + (b-t)^{r+1}}{r+1},$$

then

$$C_r\left(a,b\right) \le \int_a^b \left[\int_a^t \left(\frac{t-s}{2}\right)^r ds + \int_t^b \left(\frac{s-t}{2}\right)^r ds\right] dt$$

$$\le \frac{1}{2^r} \int_a^b \frac{(t-a)^{r+1} + (b-t)^{r+1}}{r+1} dt$$

$$= \frac{(b-a)^{r+2}}{2^{r-1}\left(r+1\right)\left(r+2\right)},$$

which proves the inequality (5.24).

For $r = 1$, we have

$$C_1\left(a,b\right) := \int_a^b \left[\int_a^t \sin\left(\frac{t-s}{2}\right) ds + \int_t^b \sin\left(\frac{s-t}{2}\right) ds\right] dt$$

$$= \int_a^b \left[2 - 2\cos\left(\frac{t-a}{2}\right) - 2\cos\left(\frac{b-t}{2}\right) + 2\right] dt$$

$$= 4\left(b-a\right) - 4\sin\left(\frac{b-a}{2}\right) - 4\sin\left(\frac{b-a}{2}\right)$$

$$= 8\left[\frac{b-a}{2} - \sin\left(\frac{b-a}{2}\right)\right],$$

which, by (5.23), produces the desired inequality (5.25).

Remark 19 *The case $b = 2\pi$ and $a = 0$ in the inequality (5.25) produces the simple inequality*

$$\left|D_\mathcal{C}\left(f; u, 0, 2\pi\right)\right| \le 8LK. \tag{5.29}$$

The following result for monotonic integrators also holds.

Theorem 5.3 (Dragomir 2016, [20]) *Let $f : \mathcal{C}\left(0,1\right) \to \mathbb{C}$ satisfy an H-r-Hölder-type condition on the circle $\mathcal{C}\left(0,1\right)$, where $H > 0$ and $r \in (0,1]$ are given. If $u : [a,b] \subseteq [0,2\pi] \to \mathbb{R}$ is a monotonic nondecreasing function on $[a,b]$, then*

$$\left|D_\mathcal{C}\left(f; u, a, b\right)\right| \le \frac{2^r H}{b-a} D_r\left(a,b\right) \tag{5.30}$$

$$\le \frac{H}{\left(r+1\right)\left(b-a\right)} \int_a^b \left[(t-a)^{r+1} + (b-t)^{r+1}\right] du\left(t\right)$$

$$\le \frac{H}{\left(r+1\right)} \left(b-a\right)^r \left[u\left(b\right) - u\left(a\right)\right]$$

where

$$D_r(a,b) := \int_a^b B_r(a,b;t)\, du(t) \tag{5.31}$$

and $B_r(a,b;t)$ is given by (5.13).

In particular, if f is Lipschitzian with the constant $L > 0$, then we have the simpler inequality

$$|D_{\mathbb{C}}(f;u,a,b)| \tag{5.32}$$

$$\leq \frac{8L}{b-a} \int_a^b \left[\sin^2\left(\frac{t-a}{4}\right) + \sin^2\left(\frac{b-t}{4}\right)\right] du(t)$$

$$\leq \frac{L}{2}(b-a)[u(b) - u(a)].$$

Proof. It is well known that if $p : [c,d] \to \mathbb{C}$ is a continuous function and $v : [c,d] \to \mathbb{R}$ is monotonic nondecreasing on $[c,d]$, then the Riemann–Stieltjes integral $\int_c^d p(t)\, dv(t)$ exists and the following inequality holds

$$\left|\int_c^d p(t)\, dv(t)\right| \leq \int_c^d |p(t)|\, dv(t). \tag{5.33}$$

Utilizing this property and the identity (5.16) we have

$$|D_{\mathbb{C}}(f;u,a,b)| \tag{5.34}$$

$$= \frac{1}{b-a}\left|\int_a^b \left(\int_a^b \left[f\left(e^{it}\right) - f\left(e^{is}\right)\right] ds\right) du(t)\right|$$

$$\leq \frac{1}{b-a} \int_a^b \left|\left(\int_a^b \left[f\left(e^{it}\right) - f\left(e^{is}\right)\right] ds\right)\right| du(t)$$

$$\leq \frac{1}{b-a} \int_a^b \left(\int_a^b \left|\left(f\left(e^{it}\right) - f\left(e^{is}\right)\right)\right| ds\right) du(t)$$

$$\leq \frac{H}{b-a} \int_a^b \left(\int_a^b \left|e^{is} - e^{it}\right|^r ds\right) du(t)$$

$$= \frac{2^r H}{b-a} \int_a^b \left[\int_a^t \sin^r\left(\frac{t-s}{2}\right) ds + \int_t^b \sin^r\left(\frac{s-t}{2}\right) ds\right] du(t).$$

We also have that

$$\int_a^b \left[\int_a^t \sin^r \left(\frac{t-s}{2} \right) ds + \int_t^b \sin^r \left(\frac{s-t}{2} \right) ds \right] du(t)$$

$$\leq \int_a^b \left[\int_a^t \left(\frac{t-s}{2} \right)^r ds + \int_t^b \left(\frac{s-t}{2} \right)^r ds \right] du(t)$$

$$= \frac{1}{2^r} \int_a^b \frac{(t-a)^{r+1} + (b-t)^{r+1}}{r+1} du(t)$$

$$= \frac{1}{2^r(r+1)} \int_a^b \left[(t-a)^{r+1} + (b-t)^{r+1} \right] du(t)$$

and the first part of the inequality (5.30) is proved.

Since

$$\max_{t \in [a,b]} \left[(t-a)^{r+1} + (b-t)^{r+1} \right] = (b-a)^{r+1}$$

then the last part of (5.30) is also proved.

For $r = 1$ we have

$$D_1(a,b) \quad : \quad = \int_a^b B_1(a,b;t) \, du(t)$$

$$= 4 \int_a^b \left[\sin^2 \left(\frac{t-a}{4} \right) + \sin^2 \left(\frac{b-t}{4} \right) \right] du(t)$$

and the inequality (5.32) is obtained.

Remark 20 *The case $a = 0, b = 2\pi$ can be stated as*

$$|D_C(f;u,0,2\pi)| \leq \frac{4L}{\pi} [u(2\pi) - u(0)]. \tag{5.35}$$

Indeed, by (5.32) we have

$$|D_C(f;u,0,2\pi)| \leq \frac{8L}{2\pi} \int_0^{2\pi} \left[\sin^2 \left(\frac{t}{4} \right) + \sin^2 \left(\frac{2\pi - t}{4} \right) \right] du(t)$$

$$= \frac{4L}{\pi} \int_0^{2\pi} \left[\sin^2 \left(\frac{t}{4} \right) + \sin^2 \left(\frac{\pi}{2} - \frac{t}{4} \right) \right] du(t)$$

$$= \frac{4L}{\pi} \int_0^{2\pi} \left[\sin^2 \left(\frac{t}{4} \right) + \cos^2 \left(\frac{t}{4} \right) \right] du(t)$$

$$= \frac{4L}{\pi} [u(2\pi) - u(0)].$$

5.3 APPLICATIONS FOR FUNCTIONS OF UNITARY OPERATORS

We have the following vector inequality for functions of unitary operators.

Theorem 5.4 (Dragomir 2016, [20]) *Assume that* $f : \mathcal{C}(0,1) \to \mathbb{C}$ *satisfies an* L-*Lipschitz-type condition on the circle* $\mathcal{C}(0,1)$, *where* $L > 0$ *is given. If the operator* $U : H \to H$ *on the Hilbert space* H *is unitary and* $\{E_\lambda\}_{\lambda \in [0,2\pi]}$ *is its spectral family, then*

$$\left| \langle f(U) x, y \rangle - \frac{1}{2\pi} \int_0^{2\pi} f\left(e^{it}\right) dt \cdot \langle x, y \rangle \right| \qquad (5.36)$$

$$\leq \frac{4L}{\pi} \bigvee_0^{2\pi} \left(\left\langle E_{(\cdot)} x, y \right\rangle \right) \leq \frac{4L}{\pi} \|x\| \|y\|$$

for any $x, y \in H$.

Proof. For given $x, y \in H$, define the function $u(\lambda) := \langle E_\lambda x, y \rangle, \lambda \in [0, 2\pi]$. We know that u is of bounded variation and

$$\bigvee_0^{2\pi} (u) =: \bigvee_0^{2\pi} \left(\left\langle E_{(\cdot)} x, y \right\rangle \right) \leq \|x\| \|y\|. \qquad (5.37)$$

On making use of the spectral representation theorem and the inequality (5.37) we deduce the desired result (5.36).

Remark 21 *Consider the function* $f : \mathcal{C}(0,1) \to \mathbb{C}$, $f_a(z) = \frac{1}{1-az}$ *with* a *real and* $0 < |a| < 1$. *We know that this function is Lipschitzian with the constant* $L = \frac{|a|}{(1-|a|)^2}$. *Since* $|ae^{it}| = |a| < 1$, *then*

$$\int_0^{2\pi} f\left(e^{it}\right) dt = \int_0^{2\pi} \frac{1}{1 - ae^{it}} dt = \int_0^{2\pi} \sum_{n=0}^{\infty} \left(ae^{it}\right)^n dt$$

$$= \sum_{n=0}^{\infty} a^n \int_0^{2\pi} \left(e^{it}\right)^n dt = \int_0^{2\pi} dt = 2\pi,$$

since for any natural number $n \geq 1$ we have $\int_0^{2\pi} \left(e^{it}\right)^n dt = 0$.
Applying the inequality (5.36) we have

$$\left| \left\langle (1_H - aU)^{-1} x, y \right\rangle - \langle x, y \rangle \right| \qquad (5.38)$$

$$\leq \frac{4\,|a|}{\pi\,(1 - |a|)^2} \bigvee_0^{2\pi} \left(\left\langle E_{(\cdot)} x, y \right\rangle \right) \leq \frac{4\,|a|}{\pi\,(1 - |a|)^2} \|x\| \, \|y\|$$

for any $x, y \in H$.

We consider the following partition of the interval $[a, b]$

$$\Delta_n : a = x_0 < x_1 < \ldots < x_{n-1} < x_n = b.$$

Define $h_k := x_{k+1} - x_k$, $0 \leq k \leq n-1$ and $\nu\left(\Delta_n\right) = \max\{h_k : 0 \leq k \leq n-1\}$ the norm of the partition Δ_n.

For the continuous function $f : \mathcal{C}(0,1) \to \mathbb{C}$ and the function $u : [a, b] \subseteq [0, 2\pi] \to \mathbb{C}$ of bounded variation on $[a, b]$, define the quadrature rule

$$D_n\left(f, u, \Delta_n\right) := \sum_{k=0}^{n-1} \frac{u\left(x_{k+1}\right) - u\left(x_k\right)}{x_{k+1} - x_k} \int_{x_k}^{x_{k+1}} f\left(e^{it}\right) dt \qquad (5.39)$$

and the remainder $R_n\left(f, u, \Delta_n\right)$ in approximating the Riemann–Stieltjes integral $\int_a^b f\left(e^{it}\right) du\left(t\right)$ by $D_n\left(f, u, \Delta_n\right)$. Then we have

$$\int_a^b f\left(e^{it}\right) du\left(t\right) = D_n\left(f, u, \Delta_n\right) + R_n\left(f, u, \Delta_n\right). \qquad (5.40)$$

The following result provides *a priori* bounds for $R_n\left(f, u, \Delta_n\right)$ in several instances of f and u as above.

Proposition 5.5 (Dragomir 2016, [20]) *Assume that $f : \mathcal{C}(0,1) \to \mathbb{C}$ satisfies the following Lipschitz-type condition*

$$|f(z) - f(w)| \leq L\,|z - w|$$

for any $w, z \in \mathcal{C}(0,1)$, where $L > 0$ is given.
If $[a, b] \subseteq [0, 2\pi]$ and the function $u : [a, b] \to \mathbb{C}$ is of

bounded variation on $[a, b]$, *then for any partition* $\Delta_n : a = x_0 < x_1 < ... < x_{n-1} < x_n = b$ *with the norm* $\nu(\Delta_n) < 2\pi$ *we have the error bound*

$$|R_n(f, u, \Delta_n)| \leq 8L \sum_{k=0}^{n-1} \frac{1}{x_{k+1} - x_k} \sin^2\left(\frac{x_{k+1} - x_k}{4}\right) \bigvee_{x_k}^{x_{k+1}}(u)$$
(5.41)

$$\leq \frac{1}{2} L\nu(\Delta_n) \bigvee_a^b(u).$$

Proof. Since $\nu(\Delta_n) < 2\pi$, then on writing inequality (5.14) on each interval $[x_k, x_{k+1}]$, where $0 \leq k \leq n - 1$, we have

$$\left| \int_{x_k}^{x_{k+1}} f\left(e^{it}\right) du(t) - \frac{u(x_{k+1}) - u(x_k)}{x_{k+1} - x_k} \int_{x_k}^{x_{k+1}} f\left(e^{it}\right) dt \right|$$

$$\leq \frac{8L}{x_{k+1} - x_k} \sin^2\left(\frac{x_{k+1} - x_k}{4}\right) \bigvee_{x_k}^{x_{k+1}}(u).$$

Utilizing the generalized triangle inequality we then have

$$|R_n (f, u, \Delta_n)|$$

$$= \left| \sum_{k=0}^{n-1} \left[\int_{x_k}^{x_{k+1}} f\left(e^{it}\right) du\,(t) - \frac{u\,(x_{k+1}) - u\,(x_k)}{x_{k+1} - x_k} \right. \right.$$

$$\times \left. \left. \int_{x_k}^{x_{k+1}} f\left(e^{it}\right) dt \right] \right|$$

$$\leq \sum_{k=0}^{n-1} \left| \left[\int_{x_k}^{x_{k+1}} f\left(e^{it}\right) du\,(t) - \frac{u\,(x_{k+1}) - u\,(x_k)}{x_{k+1} - x_k} \right. \right.$$

$$\times \left. \left. \int_{x_k}^{x_{k+1}} f\left(e^{it}\right) dt \right] \right|$$

$$\leq \sum_{k=0}^{n-1} \frac{8L}{x_{k+1} - x_k} \sin^2 \left(\frac{x_{k+1} - x_k}{4} \right) \bigvee_{x_k}^{x_{k+1}} (u)$$

$$\leq 8L \max_{0 \leq k \leq n-1} \left\{ \frac{1}{x_{k+1} - x_k} \sin^2 \left(\frac{x_{k+1} - x_k}{4} \right) \right\} \sum_{k=0}^{n-1} \bigvee_{x_k}^{x_{k+1}} (u)$$

$$= 8L \max_{0 \leq k \leq n-1} \left\{ \frac{1}{x_{k+1} - x_k} \sin^2 \left(\frac{x_{k+1} - x_k}{4} \right) \right\} \bigvee_{a}^{b} (u) .$$

Since

$$\frac{1}{x_{k+1} - x_k} \sin^2 \left(\frac{x_{k+1} - x_k}{4} \right) \leq \frac{1}{16} (x_{k+1} - x_k)$$

then

$$\max_{0 \leq k \leq n-1} \left\{ \frac{1}{x_{k+1} - x_k} \sin^2 \left(\frac{x_{k+1} - x_k}{4} \right) \right\} \leq \frac{1}{16} \nu\,(\Delta_n)$$

and the last part of (5.41) also holds.

Remark 22 *The above proposition has some particular cases of interest. If we take for instance $a = 0$, $x_1 = \pi$ and $b = 2\pi$, then we have from (5.41) that*

$$\left| \int_0^{2\pi} f\left(e^{it}\right) du\,(t) - \frac{u\,(\pi) - u\,(0)}{\pi} \int_0^{\pi} f\left(e^{it}\right) dt \right.$$

$$\left. - \frac{u\,(2\pi) - u\,(\pi)}{\pi} \int_\pi^{2\pi} f\left(e^{it}\right) dt \right| \leq \frac{8L}{\pi} \bigvee_0^{2\pi} (u) . \quad (5.42)$$

Remark 23 *We observe that the last bound in (5.41) provides a simple way to choose a division such that the accuracy in approximation is better than a given small $\varepsilon > 0$. Indeed, if we want*

$$\frac{1}{2} L \nu (\Delta_n) \bigvee_a^b (u) \leq \varepsilon$$

then we need to take Δ_n such that

$$\nu (\Delta_n) \leq \frac{2\varepsilon}{\bigvee_a^b (u) L}.$$

The above proposition can also be utilized to approximate functions of unitary operators as follows.

We consider the following partition of the interval $[0, 2\pi]$

$$\Gamma_n : 0 = \lambda_0 < \lambda_1 < ... < \lambda_{n-1} < \lambda_n = 2\pi$$

where $0 \leq k \leq n - 1$.

If U is a unitary operator on the Hilbert space H and $\{E_\lambda\}_{\lambda \in [0,2\pi]}$, the spectral family of U, then we can introduce the following sums:

$$D_n (f, U, \Gamma_n; x, y) \tag{5.43}$$

$$:= \sum_{k=0}^{n-1} \frac{1}{\lambda_{k+1} - \lambda_k} \int_{\lambda_k}^{\lambda_{k+1}} f\left(e^{it}\right) dt \cdot \left\langle (E_{\lambda_{k+1}} - E_{\lambda_k}) x, y \right\rangle.$$

Corollary 5.6 (Dragomir 2016, [20]) *Assume that $f : \mathcal{C}(0, 1) \to \mathbb{C}$ satisfies the following Lipschitz-type condition*

$$|f (z) - f (w)| \leq L |z - w|$$

for any $w, z \in \mathcal{C}(0, 1)$, where $L > 0$ is given. Assume also that U is a unitary operator on the Hilbert space H and $\{E_\lambda\}_{\lambda \in [0,2\pi]}$ is the spectral family of U.

If Γ_n is a partition of the interval $[0, 2\pi]$ with $\nu (\Gamma_n) < 2\pi$ then we have the representation

$$\langle f (U) x, y \rangle = D_n (f, U, \Gamma_n; x, y) + R_n (f, U, \Gamma_n; x, y) \tag{5.44}$$

with the error $R_n \left(f, U, \Delta_n; x, y \right)$ *satisfying the bounds*

$$\left| R_n \left(f, U, \Gamma_n; x, y \right) \right| \qquad (5.45)$$

$$\leq 8L \sum_{k=0}^{n-1} \frac{1}{\lambda_{k+1} - \lambda_k} \sin^2 \left(\frac{\lambda_{k+1} - \lambda_k}{4} \right) \bigvee_{\lambda_k}^{\lambda_{k+1}} \left(\left\langle E_{(\cdot)} x, y \right\rangle \right)$$

$$\leq \frac{1}{2} L \nu \left(\Gamma_n \right) \bigvee_{0}^{2\pi} \left(\left\langle E_{(\cdot)} x, y \right\rangle \right) \leq \frac{1}{2} L \nu \left(\Gamma_n \right) \|x\| \|y\|$$

for any $x, y \in H$.

Remark 24 *Consider the exponential mean*

$$E_z \left(p, q \right) := \frac{\exp \left(pz \right) - \exp \left(qz \right)}{p - q}$$

defined for complex numbers z *and the real numbers* p, q *with* $p \neq q$.

For the function $f \left(z \right) = z^m$ *with* m *an integer we have*

$$\int_q^p f \left(e^{it} \right) dt = \int_q^p e^{imt} dt = \frac{1}{im} \left(e^{imp} - e^{imq} \right)$$

$$= \frac{1}{im} \left(p - q \right) E_{e^{im}} \left(p, q \right).$$

For a partition Γ_n *as above, define the sum*

$$P_n \left(U, \Gamma_n; x, y \right) := \frac{1}{im} \sum_{k=0}^{n-1} E_{e^{im}} \left(\lambda_{k+1}, \lambda_k \right) \left\langle \left(E_{\lambda_{k+1}} - E_{\lambda_k} \right) x, y \right\rangle.$$

$$(5.46)$$

We can approximate the power m *of a unitary operator as follows:*

$$\left\langle U^m x, y \right\rangle = P_n \left(U, \Gamma_n; x, y \right) + T_n \left(U, \Gamma_n; x, y \right) \qquad (5.47)$$

where the error $T_n(U, \Gamma_n; x, y)$ *satisfies the bounds*

$$|T_n(U, \Gamma_n; x, y)| \tag{5.48}$$

$$\leq 8\,|m| \sum_{k=0}^{n-1} \frac{1}{\lambda_{k+1} - \lambda_k} \sin^2\left(\frac{\lambda_{k+1} - \lambda_k}{4}\right) \bigvee_{\lambda_k}^{\lambda_{k+1}} \left(\left\langle E_{(\cdot)}x, y\right\rangle\right)$$

$$\leq \frac{1}{2}\,|m|\,\nu(\Gamma_n) \bigvee_{0}^{2\pi} \left(\left\langle E_{(\cdot)}x, y\right\rangle\right) \leq \frac{1}{2}\,|m|\,\nu(\Gamma_n)\,\|x\|\,\|y\|$$

for any vectors $x, y \in H$.

CHAPTER 6

Grüss-Type Inequalities

IN THIS CHAPTER we present some Riemann–Stieltjes integral inequalities of Grüss type for continuous complex-valued integrands and various classes of bounded variation integrators. Some applications for functions of unitary operators in Hilbert spaces are provided as well.

6.1 INTRODUCTION

In [13], in order to extend the Grüss inequality to the *Riemann–Stieltjes integral*, the author introduced the following *Čebyšev functional*

$$T(f, g; u) := \frac{1}{u(b) - u(a)} \int_a^b f(t) g(t) du(t) \qquad (6.1)$$

$$- \frac{1}{u(b) - u(a)} \int_a^b f(t) du(t) \cdot \frac{1}{u(b) - u(a)}$$

$$\times \int_a^b g(t) du(t),$$

where f, g are *continuous* on $[a, b]$ and u is of *bounded variation* on $[a, b]$ with $u(b) \neq u(a)$.

The following result that provides sharp bounds for the Čebyšev functional defined above was obtained in [13].

Theorem 6.1 (Dragomir 2003, [13]) *Let* $f : [a,b] \to \mathbb{R}$, $g : [a,b] \to \mathbb{C}$ *be continuous functions on* $[a,b]$ *and* $u : [a,b] \to \mathbb{C}$ *with* $u(a) \neq u(b)$. *Assume also that there exists the real constants* γ, Γ *such that*

$$\gamma \leq f(t) \leq \Gamma \quad \text{for each } t \in [a,b]. \tag{6.2}$$

a) *If* u *is of bounded variation on* $[a,b]$, *then we have the inequality*

$$|T(f,g;u)| \tag{6.3}$$

$$\leq \frac{1}{2} \cdot \frac{\Gamma - \gamma}{|u(b) - u(a)|}$$

$$\times \left\| g - \frac{1}{u(b) - u(a)} \int_a^b g(s)\,du(s) \right\|_\infty \bigvee_a^b (u),$$

where $\bigvee_a^b (u)$ *denotes the total variation of* u *in* $[a,b]$. *The constant* $\frac{1}{2}$ *is sharp, in the sense that it cannot be replaced by a smaller quantity.*

b) *If* $u : [a,b] \to \mathbb{R}$ *is monotonic nondecreasing on* $[a,b]$, *then one has the inequality:*

$$|T(f,g;u)| \tag{6.4}$$

$$\leq \frac{1}{2} \cdot \frac{\Gamma - \gamma}{u(b) - u(a)}$$

$$\times \int_a^b \left| g(t) - \frac{1}{u(b) - u(a)} \int_a^b g(s)\,du(s) \right| du(t).$$

The constant $\frac{1}{2}$ *is sharp.*

c) *Assume that* f, g *are Riemann integrable functions on* $[a,b]$ *and* f *satisfies the condition (6.2). If* $u : [a,b] \to \mathbb{C}$ *is Lipschitzian with the constant* L, *then we have the inequality*

$$|T(f, g; u)| \qquad\qquad (6.5)$$

$$\leq \frac{1}{2} \cdot \frac{L(\Gamma - \gamma)}{|u(b) - u(a)|}$$

$$\times \int_a^b \left| g(t) - \frac{1}{u(b) - u(a)} \int_a^b g(s) \, du(s) \right| dt.$$

The constant $\frac{1}{2}$ is the best possible in (6.5).

For continuous functions $f, g : \mathcal{C}(0, 1) \to \mathbb{C}$, where $\mathcal{C}(0, 1)$ is the unit circle from $\mathbb{C}$ centered in 0 and $u : [a, b] \subseteq [0, 2\pi] \to \mathbb{C}$ is a function of bounded variation on $[a, b]$ with $u(a) \neq u(b)$, we can define the following functional as well

$$S_{\mathcal{C}}(f, g; u, a, b) := \frac{1}{u(b) - u(a)} \int_a^b f\left(e^{it}\right) g\left(e^{it}\right) du(t)$$

$$- \frac{1}{u(b) - u(a)} \int_a^b f\left(e^{it}\right) du(t) \frac{1}{u(b) - u(a)} \int_a^b g\left(e^{it}\right) du(t).$$

$$(6.6)$$

In this chapter we establish some bounds for the magnitude of $S_{\mathcal{C}}(f, g; u, a, b)$ when the *integrands* $f, g : \mathcal{C}(0, 1) \to \mathbb{C}$ satisfy some *Hölder-type conditions* on the circle $\mathcal{C}(0, 1)$ while the *integrator* u is of bounded variation.

6.2 INEQUALITIES FOR RIEMANN–STIELTJES INTE-GRAL

We say that the complex function $f : \mathcal{C}(0, 1) \to \mathbb{C}$ satisfies an *H-r-Hölder-type condition* on the circle $\mathcal{C}(0, 1)$, where $H > 0$ and $r \in (0, 1]$ are given, if

$$|f(z) - f(w)| \leq H|z - w|^r \qquad\qquad (6.7)$$

for any $w, z \in \mathcal{C}(0, 1)$.

If $r = 1$ and $L = H$, then we call it of *L-Lipschitz type*.

Consider the power function $f : \mathbb{C}\backslash\{0\} \to \mathbb{C}$, $f(z) = z^m$ where m is a nonzero integer. Then, obviously, for any z, w belonging to the unit circle $\mathcal{C}(0,1)$ we have the inequality

$$|f(z) - f(w)| \leq |m| |z - w|$$

which shows that f is Lipschitzian with constant $L = |m|$ on the circle $\mathcal{C}(0,1)$.

For $a \neq \pm 1, 0$ real numbers, consider the function $f : \mathcal{C}(0,1) \to \mathbb{C}$, $f_a(z) = \frac{1}{1-az}$. Observe that

$$|f_a(z) - f_a(w)| = \frac{|a| |z - w|}{|1 - az| |1 - aw|} \tag{6.8}$$

for any $z, w \in \mathcal{C}(0,1)$.

If $z = e^{it}$ with $t \in [0, 2\pi]$, then we have

$$
\begin{aligned}
|1 - az|^2 &= 1 - 2a\operatorname{Re}(\bar{z}) + a^2 |z|^2 = 1 - 2a\cos t + a^2 \\
&\geq 1 - 2|a| + a^2 = (1 - |a|)^2
\end{aligned}
$$

therefore

$$\frac{1}{|1 - az|} \leq \frac{1}{|1 - |a||} \quad \text{and} \quad \frac{1}{|1 - aw|} \leq \frac{1}{|1 - |a||} \tag{6.9}$$

for any $z, w \in \mathcal{C}(0,1)$.

Utilizing (6.8) and (6.9) we deduce

$$|f_a(z) - f_a(w)| \leq \frac{|a|}{(1 - |a|)^2} |z - w| \tag{6.10}$$

for any $z, w \in \mathcal{C}(0,1)$, showing that the function f_a is Lipschitzian with constant $L_a = \frac{|a|}{(1-|a|)^2}$ on the circle $\mathcal{C}(0,1)$.

The following result holds:

Theorem 6.2 (Dragomir 2015, [18]) *Assume that* $f : \mathcal{C}(0,1) \to \mathbb{C}$ *is of* H-r-*Hölder's type and* $g : \mathcal{C}(0,1) \to \mathbb{C}$ *is of* K-q-*Hölder's type. If* $u : [a,b] \subseteq [0, 2\pi] \to \mathbb{C}$ *is a function of bounded variation with* $u(a) \neq u(b)$, *then*

$$|S_{\mathcal{C}}(f, g; u, a, b)| \leq HK B_{r,q}(a, b) \left[\frac{1}{|u(b) - u(a)|} \bigvee_a^b (u) \right]^2 \tag{6.11}$$

where

$$B_{r,q}(a,b) := 2^{r+q-1} \max_{(s,t)\in[a,b]^2} \left| \sin\left(\frac{s-t}{2}\right) \right|^{r+q}. \qquad (6.12)$$

Proof. We have the following identity

$$S_{\mathcal{C}}(f,g;u,a,b) = \frac{1}{2[u(b)-u(a)]^2}$$

$$\times \int_a^b \left(\int_a^b \left[f\left(e^{it}\right) - f\left(e^{is}\right) \right] \left[g\left(e^{it}\right) - g\left(e^{it}\right) \right] du(s) \right) du(t).$$

$$(6.13)$$

It is known that if $p : [c,d] \to \mathbb{C}$ is a continuous function and $v : [c,d] \to \mathbb{C}$ is of bounded variation, then the Riemann–Stieltjes integral $\int_c^d p(t)\, dv(t)$ exists and the following inequality holds

$$\left| \int_c^d p(t)\, dv(t) \right| \le \max_{t\in[c,d]} |p(t)| \bigvee_c^d (v). \qquad (6.14)$$

Applying this property twice, we have

$$|S_{\mathcal{C}}(f,g;u,a,b)| \qquad\qquad\qquad\qquad\qquad (6.15)$$

$$= \frac{1}{2|u(b)-u(a)|^2}$$

$$\times \left| \int_a^b \left(\int_a^b \left[f\left(e^{it}\right) - f\left(e^{is}\right) \right] \left[g\left(e^{it}\right) - g\left(e^{it}\right) \right] du(s) \right) du(t) \right|$$

$$\le \frac{1}{2|u(b)-u(a)|^2}$$

$$\times \max_{t\in[a,b]} \left| \int_a^b \left[f\left(e^{it}\right) - f\left(e^{is}\right) \right] \left[g\left(e^{it}\right) - g\left(e^{it}\right) \right] du(s) \right| \bigvee_a^b (u)$$

$$\le \frac{1}{2|u(b)-u(a)|^2}$$

$$\times \max_{(t,s)\in[a,b]^2} \left| \left[f\left(e^{it}\right) - f\left(e^{is}\right) \right] \left[g\left(e^{it}\right) - g\left(e^{it}\right) \right] \right| \left[\bigvee_a^b (u) \right]^2.$$

Utilizing the properties of f and g we have

$$\left|\left[f\left(e^{it}\right) - f\left(e^{is}\right)\right]\left[g\left(e^{it}\right) - g\left(e^{it}\right)\right]\right| \le HK\left|e^{is} - e^{it}\right|^{r+q} \tag{6.16}$$

for any $s, t \in [a, b]$.

Since

$$\begin{aligned}\left|e^{is} - e^{it}\right|^2 &= \left|e^{is}\right|^2 - 2\operatorname{Re}\left(e^{i(s-t)}\right) + \left|e^{it}\right|^2\\ &= 2 - 2\cos(s - t) = 4\sin^2\left(\frac{s - t}{2}\right)\end{aligned}$$

for any $t, s \in \mathbb{R}$, then

$$\left|e^{is} - e^{it}\right|^{r+q} = 2^{r+q}\left|\sin\left(\frac{s - t}{2}\right)\right|^{r+q} \tag{6.17}$$

for any $t, s \in \mathbb{R}$.

Utilizing (6.15) and (6.17) we deduce the desired result (6.11).

Remark 25 *If $b = 2\pi$ and $a = 0$ then obviously there are $s, t \in [0, 2\pi]$ such that $s - t = \pi$ showing that*

$$\max_{(s,t)\in[0,2\pi]^2}\left|\sin\left(\frac{s - t}{2}\right)\right|^{r+q} = 1.$$

In this situation we have

$$\left|S_C\left(f, g; u, 0, 2\pi\right)\right| \le 2^{r+q-1}HK\left[\frac{1}{\left|u\left(2\pi\right) - u\left(0\right)\right|}\bigvee_0^{2\pi}\left(u\right)\right]^2. \tag{6.18}$$

Moreover, if f and g are Lipschitzian with constants L and N, respectively, the inequality (6.18) becomes

$$\left|S_C\left(f, g; u, 0, 2\pi\right)\right| \le 2LN\left[\frac{1}{\left|u\left(2\pi\right) - u\left(0\right)\right|}\bigvee_0^{2\pi}\left(u\right)\right]^2. \tag{6.19}$$

Remark 26 *For intervals smaller than π, i.e., $0 < b - a \le \pi$, then for all $t, s \in [a, b] \subseteq [0, 2\pi]$ we have $\frac{1}{2}\left|t - s\right| \le$*

$\frac{1}{2}(b-a) \leq \frac{\pi}{2}$. *Since the function* $\sin$ *is increasing on* $\left[0, \frac{\pi}{2}\right]$, *then we have successively that*

$$\max_{(t,s)\in[a,b]^2} \left| \sin\left(\frac{s-t}{2}\right) \right| = \sin\left(\max_{(t,s)\in[a,b]^2} \frac{1}{2}|t-s| \right) = \sin\left(\frac{b-a}{2}\right).$$
(6.20)

In this case we get the inequality

$$|S_{\mathcal{C}}(f,g;u,a,b)| \leq HKB_{r,q}(a,b)\left[\frac{1}{|u(b)-u(a)|} \bigvee_a^b (u) \right]^2$$
(6.21)

where

$$B_{r,q}(a,b) := 2^{r+q-1} \left| \sin\left(\frac{b-a}{2}\right) \right|^{r+q}.$$
(6.22)

Moreover, if f *and* g *are Lipschitzian with constants* L *and* N, *respectively, then*

$$B(a,b) := B_{1,1}(a,b) = 2\sin^2\left(\frac{b-a}{2}\right)$$

and the inequality (6.21) becomes

$$|S_{\mathcal{C}}(f,g;u,a,b)| \leq 2LN\sin^2\left(\frac{b-a}{2}\right)\left[\frac{1}{|u(b)-u(a)|} \bigvee_a^b (u) \right]^2.$$
(6.23)

We also have:

Theorem 6.3 (Dragomir 2015, [18]) *Assume that* $f : \mathcal{C}(0,1) \to \mathbb{C}$ *is of* H-r-*Hölder's type and* $g : \mathcal{C}(0,1) \to \mathbb{C}$ *is of* K-q-*Hölder's type. If* $u : [a,b] \subseteq [0,2\pi] \to \mathbb{C}$ *is an* M-*Lipschitzian function with* $u(a) \neq u(b)$, *then*

$$|S_{\mathcal{C}}(f,g;u,a,b)| \leq 2^{p+q-1} \frac{M^2HK}{|u(b)-u(a)|^2} C_{p,q}(a,b)$$
(6.24)

where

$$C_{r,q}(a,b) \quad : \quad = \int_a^b \int_a^b \left| \sin\left(\frac{s-t}{2}\right) \right|^{r+q} ds\, dt$$
(6.25)

$$\leq \frac{1}{2^{r+q-1}(r+q+1)(r+q+2)}(b-a)^{r+q+2}.$$

Proof. It is well known that if $p : [a, b] \to \mathbb{C}$ is a Riemann integrable function and $v : [a, b] \to \mathbb{C}$ is Lipschitzian with the constant $M > 0$, then the Riemann–Stieltjes integral $\int_a^b p(t) \, dv(t)$ exists and the following inequality holds

$$\left| \int_a^b p(t) \, dv(t) \right| \leq M \int_a^b |p(t)| \, dt. \qquad (6.26)$$

Utilizing this property and the identity (6.13) we have

$$|S_{\mathcal{C}}(f, g; u, a, b)| \qquad (6.27)$$

$$= \frac{1}{2 |u(b) - u(a)|^2}$$

$$\times \left| \int_a^b \left(\int_a^b \left[f\left(e^{it}\right) - f\left(e^{is}\right) \right] \left[g\left(e^{it}\right) - g\left(e^{it}\right) \right] du(s) \right) \right.$$

$$\left. \times \, du(t) \right|$$

$$\leq \frac{M}{2 |u(b) - u(a)|^2}$$

$$\times \int_a^b \left| \int_a^b \left[f\left(e^{it}\right) - f\left(e^{is}\right) \right] \left[g\left(e^{it}\right) - g\left(e^{it}\right) \right] du(s) \right| dt$$

$$\leq \frac{M^2}{2 |u(b) - u(a)|^2}$$

$$\times \int_a^b \int_a^b \left| \left[f\left(e^{it}\right) - f\left(e^{is}\right) \right] \left[g\left(e^{it}\right) - g\left(e^{it}\right) \right] \right| ds dt.$$

Utilizing the properties of f and g we have

$$\left| \left[f\left(e^{it}\right) - f\left(e^{is}\right) \right] \left[g\left(e^{it}\right) - g\left(e^{it}\right) \right] \right| \leq HK \left| e^{is} - e^{it} \right|^{r+q}$$

for any $s, t \in [a, b]$, which implies that

$$\int_a^b \int_a^b \left| \left[f\left(e^{it}\right) - f\left(e^{is}\right) \right] \left[g\left(e^{it}\right) - g\left(e^{it}\right) \right] \right| ds dt$$

$$\leq HK \int_a^b \int_a^b \left| e^{is} - e^{it} \right|^{r+q} ds dt$$

$$= 2^{r+q} HK \int_a^b \int_a^b \left| \sin\left(\frac{s-t}{2}\right) \right|^{r+q} ds dt.$$

Utilizing the well-known inequality

$$|\sin x| \le |x| \text{ for any } x \in \mathbb{R}$$

we have

$$
\int_a^b \int_a^b \left| \sin \left(\frac{s-t}{2} \right) \right|^{r+q} ds dt
$$

$$
\le \frac{1}{2^{r+q}} \int_a^b \int_a^b |s-t|^{r+q} ds dt
$$

$$
= \frac{1}{2^{r+q}} \int_a^b \left[\int_a^t (t-s)^{r+q} ds + \int_t^b (s-t)^{r+q} ds \right] dt
$$

$$
= \frac{1}{2^{r+q}} \int_a^b \frac{(t-a)^{r+q+1} + (b-t)^{r+q+1}}{r+q+1} dt
$$

$$
= \frac{2 (b-a)^{r+q+2}}{2^{r+q} (r+q+1)(r+q+2)}
$$

$$
= \frac{1}{2^{r+q-1} (r+q+1)(r+q+2)} (b-a)^{r+q+2}
$$

and the bound in (6.25) is proved.

The case of Lipschitzian integrators is of importance and can be stated as follows:

Corollary 6.4 (Dragomir 2015, [18]) *Assume that* u : $[a,b] \subseteq [0, 2\pi] \to \mathbb{C}$ *is an* M-*Lipschitzian function with* $u(a) \ne u(b)$. *If* f *and* g *are Lipschitzian with constants* L *and* N, *respectively, then*

$$
|S_{\mathcal{C}}(f, g; u, a, b)| \le \frac{4M^2 N L}{|u(b) - u(a)|^2} \left[\left(\frac{b-a}{2} \right)^2 - \sin^2 \left(\frac{b-a}{2} \right) \right].
$$
(6.28)

Proof. We have to calculate

$$
C_{1,1}(a,b) = \int_a^b \int_a^b \sin^2 \left(\frac{s-t}{2} \right) ds dt
$$

$$
= \int_a^b \int_a^b \frac{1 - \cos(s-t)}{2} ds dt
$$

$$
= \frac{1}{2} \left[(b-a)^2 - \int_a^b [\sin(b-t) - \sin(a-t)] dt \right].
$$

Since

$$\int_a^b \sin(b-t)\, dt = 1 - \cos(b-a)$$

and

$$\int_a^b \sin(a-t)\, dt = \cos(b-a) - 1$$

then

$$
\begin{aligned}
C_{1,1}(a,b) &= \frac{1}{2}\left[(b-a)^2 - 2(1-\cos(b-a))\right] \\
&= \frac{1}{2}\left[(b-a)^2 - 4\sin^2\left(\frac{b-a}{2}\right)\right] \\
&= 2\left[\left(\frac{b-a}{2}\right)^2 - \sin^2\left(\frac{b-a}{2}\right)\right]
\end{aligned}
$$

and the inequality (6.28) then follows from (6.24).

The case of monotonic nondecreasing integrators is as follows:

Theorem 6.5 (Dragomir 2015, [18]) *Assume that $f : \mathcal{C}(0,1) \to \mathbb{C}$ is of H-r-Hölder's type and $g : \mathcal{C}(0,1) \to \mathbb{C}$ is of K-q-Hölder's type. If $u : [a,b] \subseteq [0,2\pi] \to \mathbb{R}$ is a monotonic nondecreasing function with $u(a) < u(b)$, then*

$$|S_{\mathcal{C}}(f,g;u,a,b)| \leq \frac{2^{p+q-1}HK}{[u(b)-u(a)]^2} D_{r,q}(a,b) \qquad (6.29)$$

where

$$D_{r,q}(a,b) := \int_a^b \int_a^b \left|\sin\left(\frac{s-t}{2}\right)\right|^{r+q} du(s)\, du(t) \qquad (6.30)$$

$$\leq \max_{s,t\in[a,b]^2} \left|\sin\left(\frac{s-t}{2}\right)\right|^{r+q} [u(b)-u(a)]^2.$$

Proof. It is well known that if $p : [a,b] \to \mathbb{C}$ is a continuous function and $v : [a,b] \to \mathbb{R}$ is monotonic nondecreasing on $[a,b]$, then the Riemann–Stieltjes integral $\int_a^b p(t)\, dv(t)$ exists and the following inequality holds

$$\left|\int_a^b p(t)\, dv(t)\right| \leq \int_a^b |p(t)|\, dv(t). \qquad (6.31)$$

Utilizing this property and the identity (6.13) we have

$$|S_C(f, g; u, a, b)| \qquad\qquad (6.32)$$

$$= \frac{1}{2[u(b) - u(a)]^2}$$

$$\times \left| \int_a^b \left(\int_a^b [f(e^{it}) - f(e^{is})][g(e^{it}) - g(e^{it})] \, du(s) \right) du(t) \right|$$

$$\leq \frac{1}{2[u(b) - u(a)]^2}$$

$$\times \int_a^b \left| \int_a^b [f(e^{it}) - f(e^{is})][g(e^{it}) - g(e^{it})] \, du(s) \right| du(t)$$

$$\leq \frac{1}{2[u(b) - u(a)]^2}$$

$$\times \int_a^b \int_a^b \left| [f(e^{it}) - f(e^{is})][g(e^{it}) - g(e^{it})] \right| du(s) \, du(t).$$

Since

$$\left| [f(e^{it}) - f(e^{is})][g(e^{it}) - g(e^{it})] \right| \leq HK \left| e^{is} - e^{it} \right|^{r+q}$$

for any $s, t \in [a, b]$, then

$$\int_a^b \int_a^b \left| [f(e^{it}) - f(e^{is})][g(e^{it}) - g(e^{it})] \right| du(s) \, du(t)$$

$$\leq HK \int_a^b \int_a^b \left| e^{is} - e^{it} \right|^{r+q} du(s) \, du(t)$$

$$= 2^{r+q} HK \int_a^b \int_a^b \left| \sin\left(\frac{s-t}{2} \right) \right|^{r+q} du(s) \, du(t)$$

and the inequality (6.29) is proved.

The bound (6.30) for $D_{r,q}(a, b)$ is obvious.

The Lipschitzian case is of interest due to many examples that can be provided as follows:

Corollary 6.6 (Dragomir 2015, [18]) *If f and g are Lipschitzian with constants L and N, respectively, and $u : [a, b] \subseteq$*

$[0, 2\pi] \rightarrow \mathbb{R}$ *is a monotonic nondecreasing function with* $u(a) < u(b)$, *then*

$$|S_{\mathcal{C}}(f, g; u, a, b)| \qquad (6.33)$$

$$\leq \frac{LN}{[u(b) - u(a)]^2}$$

$$\times \left[[u(b) - u(a)]^2 - \left(\int_a^b \cos s \, du(s) \right)^2 - \left(\int_a^b \sin s \, du(s) \right)^2 \right].$$

Proof. We have to calculate

$$\begin{aligned} D_{1,1}(a, b) &= \int_a^b \int_a^b \sin^2 \left(\frac{s - t}{2} \right) du(s) \, du(t) \\ &= \int_a^b \int_a^b \frac{1 - \cos(s - t)}{2} du(s) \, du(t) \\ &= \frac{1}{2} \left[(u(b) - u(a))^2 - J(a, b) \right] \end{aligned}$$

where

$$J(a, b) := \int_a^b \int_a^b \cos(s - t) \, du(s) \, du(t).$$

Since

$$\cos(s - t) = \cos s \cos t + \sin s \sin t$$

then

$$J(a, b) = \left(\int_a^b \cos s \, du(s) \right)^2 + \left(\int_a^b \sin s \, du(s) \right)^2.$$

Utilizing (6.29) we deduce the desired result (6.33).

Remark 27 *Utilizing the integration by parts formula for the Riemann–Stieltjes integral, we have*

$$\int_a^b \cos s \, du(s) = u(b) \cos b - u(a) \cos a + \int_a^b u(s) \sin s \, ds$$

and

$$\int_a^b \sin s \, du(s) = u(b) \sin b - u(a) \sin a - \int_a^b u(s) \cos s \, ds.$$

If we insert these values in the right-hand side of (6.33) we can get some expressions containing only Riemann integrals. However they are complicated and will not be presented here.

6.3 APPLICATIONS FOR FUNCTIONS OF UNITARY OPERATORS

We have the following vector inequality for functions of unitary operators.

Theorem 6.7 (Dragomir 2015, [18]) *Assume that* f : $\mathcal{C}(0,1) \to \mathbb{C}$ *is of H-r-Hölder's type and g : $\mathcal{C}(0,1) \to \mathbb{C}$ is of K-q-Hölder's type. If the operator $U : H \to H$ on the Hilbert space H is unitary, then*

$$|\langle x,y\rangle \langle f(U) g(U) x,y\rangle - \langle f(U) x,y\rangle \langle g(U) x,y\rangle| \quad (6.34)$$

$$\leq 2^{r+q-1} HK \left[\bigvee_0^{2\pi} \left\langle E_{(\cdot)}x,y \right\rangle \right]^2 \leq 2^{r+q-1} HK \|x\|^2 \|y\|^2$$

for any $x, y \in H$.

In particular, if f and g are Lipschitzian with constants L and N, respectively, then

$$|\langle x,y\rangle \langle f(U) g(U) x,y\rangle - \langle f(U) x,y\rangle \langle g(U) x,y\rangle| \quad (6.35)$$

$$\leq 2LN \left[\bigvee_0^{2\pi} \left\langle E_{(\cdot)}x,y \right\rangle \right]^2 \leq 2LN \|x\|^2 \|y\|^2$$

for any $x, y \in H$.

Proof. For given $x, y \in H$, define the function $u(\lambda) := \langle E_\lambda x,y\rangle, \lambda \in [0, 2\pi]$. We know that u is of bounded variation and

$$\bigvee_0^{2\pi} (u) =: \bigvee_0^{2\pi} \left(\left\langle E_{(\cdot)}x,y \right\rangle \right) \leq \|x\| \|y\|. \quad (6.36)$$

Now, from the inequality (6.18) we have

$$
\left| \left(\langle E_{2\pi} x, y \rangle - \langle E_0 x, y \rangle \right) \int_0^{2\pi} f\left(e^{it}\right) g\left(e^{it}\right) d\langle E_t x, y \rangle \right. \quad (6.37)
$$
$$
\left. - \int_0^{2\pi} f\left(e^{it}\right) d\langle E_t x, y \rangle \int_0^{2\pi} g\left(e^{it}\right) d\langle E_t x, y \rangle \right|
$$
$$
\leq 2^{r+q-1} HK \left[\bigvee_0^{2\pi} \left(\langle E_{(\cdot)} x, y \rangle \right) \right]^2
$$

for any $x, y \in H$.

The proof is complete.

Remark 28 *If $U : H \to H$ is a unitary operator on the Hilbert space H, then for any integer m, n we have from (6.35) the power inequalities*

$$
\left| \langle x, y \rangle \langle U^{m+n} x, y \rangle - \langle U^m x, y \rangle \langle U^n x, y \rangle \right| \quad (6.38)
$$
$$
\leq 2|mn| \left[\bigvee_0^{2\pi} \langle E_{(\cdot)} x, y \rangle \right]^2 \leq 2|mn| \, \|x\|^2 \, \|y\|^2
$$

for any $x, y \in H$.

In particular, we have

$$
\left| \langle x, y \rangle \langle U^2 x, y \rangle - \langle U x, y \rangle^2 \right| \leq 2 \left[\bigvee_0^{2\pi} \langle E_{(\cdot)} x, y \rangle \right]^2 \leq 2 \, \|x\|^2 \, \|y\|^2
$$
$$
(6.39)
$$

and

$$
\left| \langle x, y \rangle^2 - \langle U x, y \rangle \langle x, U y \rangle \right| \leq 2 \left[\bigvee_0^{2\pi} \langle E_{(\cdot)} x, y \rangle \right]^2 \leq 2 \, \|x\|^2 \, \|y\|^2
$$
$$
(6.40)
$$

for any $x, y \in H$.

For $a \neq \pm 1, 0$ real numbers, consider the function $f : \mathcal{C}(0,1) \to \mathbb{C}$, $f_a(z) = \frac{1}{1-az}$. This function is Lipschitzian with the constant $L_a = \frac{|a|}{(1-|a|)^2}$ on the circle $\mathcal{C}(0,1)$.

Now, if we take $a, b \neq \pm 1, 0$ and use the inequality (6.35) then we have

$$\left| \langle x, y \rangle \left\langle (1_H - aU)^{-1} (1_H - bU)^{-1} x, y \right\rangle \right. \quad (6.41)$$
$$\left. - \left\langle (1_H - aU)^{-1} x, y \right\rangle \left\langle (1_H - bU)^{-1} x, y \right\rangle \right|$$
$$\leq \frac{2 |a| |b|}{(1 - |a|)^2 (1 - |b|)^2} \left[\bigvee_0^{2\pi} \left\langle E_{(\cdot)} x, y \right\rangle \right]^2$$
$$\leq \frac{2 |a| |b|}{(1 - |a|)^2 (1 - |b|)^2} \|x\|^2 \|y\|^2$$

for any $x, y \in H$.

In particular, we have

$$\left| \langle x, y \rangle \left\langle (1_H - aU)^{-2} x, y \right\rangle - \left\langle (1_H - aU)^{-1} x, y \right\rangle^2 \right| \quad (6.42)$$
$$\leq \frac{2 |a|^2}{(1 - |a|)^4} \left[\bigvee_0^{2\pi} \left\langle E_{(\cdot)} x, y \right\rangle \right]^2 \leq \frac{2 |a|^2}{(1 - |a|)^4} \|x\|^2 \|y\|^2$$

for any $x, y \in H$.

Theorem 6.8 (Dragomir 2015, [18]) *If f and g are Lipschitzian with constants L and N, respectively and $U : H \to H$ is a unitary operator on the Hilbert space H, then*

$$\left| \|x\|^2 \langle f(U) g(U) x, x \rangle - \langle f(U) x, x \rangle \langle g(U) x, x \rangle \right| \quad (6.43)$$
$$\leq LN \left[\|x\|^4 - \langle \mathrm{Re}(U) x, x \rangle^2 - \langle \mathrm{Im}(U) x, x \rangle^2 \right]$$
$$= LN \left[\|x\|^4 - |\langle Ux, x \rangle|^2 \right]$$

for any $x \in H$, where

$$\mathrm{Re}(U) := \frac{U + U^*}{2} \quad and \quad \mathrm{Im}(U) := \frac{U - U^*}{2i}.$$

Proof. From the inequality (6.33) we have

$$
\left| \left(\langle E_{2\pi} x, x \rangle - \langle E_0 x, x \rangle \right) \int_0^{2\pi} f\left(e^{it} \right) g\left(e^{it} \right) d \langle E_t x, x \rangle \right. \tag{6.44}
$$

$$
\left. - \int_0^{2\pi} f\left(e^{it} \right) d \langle E_t x, x \rangle \int_0^{2\pi} g\left(e^{it} \right) d \langle E_t x, x \rangle \right|
$$

$$
\leq LN \left[\left(\langle E_{2\pi} x, x \rangle - \langle E_0 x, x \rangle \right)^2 \right.
$$

$$
\left. - \left(\int_0^{2\pi} \cos t d \langle E_t x, x \rangle \right)^2 - \left(\int_0^{2\pi} \sin t d \langle E_t x, x \rangle \right)^2 \right]
$$

for any $x, y \in H$.

Since

$$
\mathrm{Re}\left(e^{it} \right) = \cos t \text{ and } \mathrm{Im}\left(e^{it} \right) = \sin t,
$$

then we have from the spectral representation theorem

$$
\left(\int_0^{2\pi} \cos t d \langle E_t x, x \rangle \right)^2 = \langle \mathrm{Re}\left(U \right) x, x \rangle^2
$$

$$
\left(\int_0^{2\pi} \sin t d \langle E_t x, x \rangle \right)^2 = \langle \mathrm{Im}\left(U \right) x, x \rangle^2
$$

and due to the fact that

$$
\begin{aligned}
\left| \langle U x, x \rangle \right|^2 &= \left| \langle \left[\mathrm{Re}\left(U \right) + i \, \mathrm{Im}\left(U \right) \right] x, x \rangle \right|^2 \\
&= \left| \langle \mathrm{Re}\left(U \right) x, x \rangle + i \, \langle \mathrm{Im}\left(U \right) x, x \rangle \right|^2 \\
&= \langle \mathrm{Re}\left(U \right) x, x \rangle^2 + \langle \mathrm{Im}\left(U \right) x, x \rangle^2
\end{aligned}
$$

we deduce from (6.44) the desired inequality (6.43).

Remark 29 *If $U : H \to H$ is a unitary operator on the Hilbert space H, then for any integer m, n we have from (6.43) the power inequalities*

$$
\left| \|x\|^2 \langle U^{n+m} x, x \rangle - \langle U^n x, x \rangle \langle U^m x, x \rangle \right| \tag{6.45}
$$

$$
\leq |mn| \left[\|x\|^4 - \left| \langle U x, x \rangle \right|^2 \right]
$$

for any $x \in H$.

In particular, we have for $n = m = 1$

$$\left| \|x\|^2 \left\langle U^2 x, x \right\rangle - \left\langle Ux, x \right\rangle^2 \right| \leq \|x\|^4 - |\langle Ux, x \rangle|^2 \qquad (6.46)$$

for any $x \in H$.

If we take $n = 1$ and $m = -n$ and take into account that

$$\left\langle U^{-n} x, x \right\rangle = \left\langle (U^n)^* x, x \right\rangle = \left\langle x, U^n x \right\rangle = \overline{\langle U^n x, x \rangle}$$

for any $x \in H$, then we get from (6.45) that

$$0 \leq \|x\|^4 - |\langle U^n x, x \rangle|^2 \leq n^2 \left[\|x\|^4 - |\langle Ux, x \rangle|^2 \right] \qquad (6.47)$$

for any $x \in H$.

Now, if we take $a, b \neq \pm 1, 0$ and use the inequality (6.43), then we get

$$\left| \|x\|^2 \left\langle (1 - aU)^{-1} (1 - bU)^{-1} x, x \right\rangle \right. \qquad (6.48)$$
$$\left. - \left\langle (1 - aU)^{-1} x, x \right\rangle \left\langle (1 - bU)^{-1} x, x \right\rangle \right|$$
$$\leq \frac{2 |a| |b|}{(1 - |a|)^2 (1 - |b|)^2} \left[\|x\|^4 - |\langle Ux, x \rangle|^2 \right]$$

for any $x \in H$.

In particular, we have

$$\left| \|x\|^2 \left\langle (1 - aU)^{-2} x, x \right\rangle - \left\langle (1 - aU)^{-1} x, x \right\rangle^2 \right| \qquad (6.49)$$
$$\leq \frac{2 |a|^2}{(1 - |a|)^4} \left[\|x\|^4 - |\langle Ux, x \rangle|^2 \right]$$

for any $x \in H$.

CHAPTER 7

Inequalities for Bounded Functions

IN THIS CHAPTER we present some Riemann–Stieltjes integral inequalities for continuous complex bounded integrands and various classes of bounded variation integrators. Some applications for functions of unitary operators in Hilbert spaces are provided as well.

7.1   SOME IDENTITIES

Let $f : \mathcal{C}(0,1) \to \mathbb{C}$ be continuous on the unit circle $\mathcal{C}(0,1) := \{z = e^{it}, \; t \in [0, 2\pi]\}$. Assume that $u : [a, b] \subset [0, 2\pi] \to \mathbb{C}$. If the Riemann–Stieltjes integral $\int_a^b f(e^{it}) \, du(t)$ exists, we write for simplicity, like in [1, p. 142] that $f(e^{i\cdot}) \in \mathcal{R}_{\mathbb{C}}(u, [a, b])$.

We start with the following simple fact:

Lemma 7.1 *Let $f, g : \mathcal{C}(0,1) \to \mathbb{C}$ be continuous on the unit circle $\mathcal{C}(0,1)$ and $v : [a, b] \to \mathbb{C}$ of bounded variation on $[a, b]$*

while $\lambda,\ \mu \in \mathbb{C}$ and $x \in [a,b]$. Then

$$\int_a^b f\left(e^{it}\right) g\left(e^{it}\right) dv(t) = \lambda \int_a^x g\left(e^{it}\right) dv(t) + \mu \int_x^b g\left(e^{it}\right) dv(t)$$

$$+ \int_a^x \left[f\left(e^{it}\right) - \lambda\right] g\left(e^{it}\right) dv(t) + \int_x^b \left[f\left(e^{it}\right) - \mu\right] g\left(e^{it}\right) dv(t)$$

$$= \mu \int_a^b g\left(e^{it}\right) dv(t) + (\lambda - \mu) \int_a^x g\left(e^{it}\right) dv(t)$$

$$+ \int_a^x \left[f\left(e^{it}\right) - \lambda\right] g\left(e^{it}\right) dv(t) + \int_x^b \left[f\left(e^{it}\right) - \mu\right] g\left(e^{it}\right) dv(t).$$

$$(7.1)$$

In particular, for $\mu = \lambda$, we have

$$\int_a^b f\left(e^{it}\right) g\left(e^{it}\right) dv(t) = \lambda \int_a^b g\left(e^{it}\right) dv(t)$$

$$+ \int_a^x \left[f\left(e^{it}\right) - \lambda\right] g\left(e^{it}\right) dv(t) + \int_x^b \left[f\left(e^{it}\right) - \lambda\right] g\left(e^{it}\right) dv(t)$$

$$= \lambda \int_a^b g\left(e^{it}\right) dv(t) + \int_a^b \left[f\left(e^{it}\right) - \lambda\right] g\left(e^{it}\right) dv(t). \quad (7.2)$$

Proof. The integrability follows by the properties of the Riemann–Stieltjes integral, see [1, Theorem 7.27]. Using the properties of the Riemann–Stieltjes integral, we have

$$\int_a^x \left[f\left(e^{it}\right) - \lambda\right] g\left(e^{it}\right) dv(t) + \int_x^b \left[f\left(e^{it}\right) - \mu\right] g\left(e^{it}\right) dv(t)$$

$$= \int_a^x f\left(e^{it}\right) g\left(e^{it}\right) dv(t) - \lambda \int_a^x g\left(e^{it}\right) dv(t)$$

$$+ \int_x^b f\left(e^{it}\right) g\left(e^{it}\right) dv(t) - \mu \int_x^b g\left(e^{it}\right) dv(t)$$

$$= \int_a^b f\left(e^{it}\right) g\left(e^{it}\right) dv(t) - \lambda \int_a^x g\left(e^{it}\right) dv(t)$$

$$- \mu \int_x^b g\left(e^{it}\right) dv(t),$$

which is equivalent to the first equality in (7.1).

The rest is obvious.

Corollary 7.2 *Let* $f : \mathcal{C}(0,1) \to \mathbb{C}$ *be continuous on the unit circle* $\mathcal{C}(0,1)$ *and* $v : [a,b] \to \mathbb{C}$ *of bounded variation on* $[a,b]$. *Then for any* $\lambda, \mu \in \mathbb{C}$ *and* $x \in [a,b]$ *we have the equality*

$$\int_a^b f\left(e^{it}\right) dv(t) = \lambda \left[v(x) - v(a)\right] + \mu \left[v(b) - v(x)\right]$$

$$+ \int_a^x \left[f\left(e^{it}\right) - \lambda\right] dv(t) + \int_x^b \left[f\left(e^{it}\right) - \mu\right] dv(t). \quad (7.3)$$

In particular, for $\mu = \lambda$, *we have*

$$\int_a^b f\left(e^{it}\right) dv(t) = \lambda \left[v(b) - v(a)\right]$$

$$+ \int_a^x \left[f\left(e^{it}\right) - \lambda\right] dv(t) + \int_x^b \left[f\left(e^{it}\right) - \lambda\right] dv(t)$$

$$= \lambda \left[v(b) - v(a)\right] + \int_a^b \left[f\left(e^{it}\right) - \lambda\right] dv(t). \quad (7.4)$$

The proof follows by Lemma 7.1 for $g\left(e^{it}\right) = 1$, $t \in [a,b]$.
If we use the equality (7.2) for $\lambda = f\left(e^{ix}\right)$, $x \in [a,b]$, $\lambda = \frac{1}{b-a} \int_a^b f\left(e^{it}\right) dt$ and $\lambda = \frac{f\left(e^{ia}\right)+f\left(e^{ib}\right)}{2}$, then we have

$$\int_a^b f\left(e^{it}\right) g\left(e^{it}\right) du(t) = f\left(e^{ix}\right) \int_a^b g\left(e^{it}\right) du(t)$$

$$- \int_a^x \left[f\left(e^{ix}\right) - f\left(e^{it}\right)\right] g\left(e^{it}\right) du(t)$$

$$+ \int_x^b \left[f\left(e^{it}\right) - f\left(e^{ix}\right)\right] g\left(e^{it}\right) du(t)$$

$$= \int_a^b \left[f\left(e^{it}\right) - f\left(e^{ix}\right)\right] g\left(e^{it}\right) du(t), \quad (7.5)$$

$$\int_a^b f\left(e^{it}\right) g\left(e^{it}\right) du(t) = \frac{1}{b-a} \int_a^b f\left(e^{it}\right) dt \int_a^b g\left(e^{it}\right) du(t)$$

$$+ \int_a^b \left[f\left(e^{it}\right) - \frac{1}{b-a} \int_a^b f\left(e^{is}\right) ds\right] g\left(e^{it}\right) du(t), \quad (7.6)$$

and

$$\int_a^b f\left(e^{it}\right) g\left(e^{it}\right) du\left(t\right) = \frac{f\left(e^{ia}\right) + f\left(e^{ib}\right)}{2} \int_a^b g\left(e^{it}\right) du\left(t\right)$$
$$+ \int_a^b \left[f\left(e^{it}\right) - \frac{f\left(e^{ia}\right) + f\left(e^{ib}\right)}{2}\right] g\left(e^{it}\right) du\left(t\right), \quad (7.7)$$

respectively.

In particular, for $g\left(e^{it}\right) = 1$, $t \in [a,b]$, we have for any $x \in [a,b]$ that

$$\int_a^b f\left(e^{it}\right) du\left(t\right) = \left[u\left(b\right) - u\left(a\right)\right] f\left(e^{ix}\right)$$
$$- \int_a^x \left[f\left(e^{ix}\right) - f\left(e^{it}\right)\right] du\left(t\right) + \int_x^b \left[f\left(e^{it}\right) - f\left(e^{ix}\right)\right] du\left(t\right)$$
$$= \left[u\left(b\right) - u\left(a\right)\right] f\left(e^{ix}\right) + \int_a^b \left[f\left(e^{it}\right) - f\left(e^{ix}\right)\right] du\left(t\right),$$
$$(7.8)$$

$$\int_a^b f\left(e^{it}\right) du\left(t\right) = \frac{u\left(b\right) - u\left(a\right)}{b - a} \int_a^b f\left(e^{it}\right) dt$$
$$+ \int_a^b \left[f\left(e^{it}\right) - \frac{1}{b - a} \int_a^b f\left(e^{it}\right) dt\right] du\left(t\right), \quad (7.9)$$

and

$$\int_a^b f\left(e^{it}\right) du\left(t\right) = \left[u\left(b\right) - u\left(a\right)\right] \frac{f\left(e^{ia}\right) + f\left(e^{ib}\right)}{2}$$
$$+ \int_a^b \left[f\left(e^{it}\right) - \frac{f\left(e^{ia}\right) + f\left(e^{ib}\right)}{2}\right] du\left(t\right), \quad (7.10)$$

respectively.

If we take $\lambda = f\left(e^{ia}\right)$ and $\mu = f\left(e^{ib}\right)$ in (7.1) we get for

$x \in [a, b]$ that

$$\int_a^b f\left(e^{it}\right) g\left(e^{it}\right) du\left(t\right)$$

$$= f\left(e^{ia}\right) \int_a^x g\left(e^{it}\right) du\left(t\right) + f\left(e^{ib}\right) \int_x^b g\left(e^{it}\right) du\left(t\right)$$

$$+ \int_a^x \left[f\left(e^{it}\right) - f\left(e^{ia}\right)\right] g\left(e^{it}\right) du\left(t\right)$$

$$+ \int_x^b \left[f\left(e^{it}\right) - f\left(e^{ib}\right)\right] g\left(e^{it}\right) du\left(t\right), \quad (7.11)$$

while for $\lambda = f\left(e^{\frac{a+x}{2}i}\right)$ and $\mu = f\left(e^{\frac{x+b}{2}i}\right)$ in (7.1) we get for $x \in [a, b]$ that

$$\int_a^b f\left(e^{it}\right) g\left(e^{it}\right) du\left(t\right)$$

$$= f\left(e^{\frac{a+x}{2}i}\right) \int_a^x g\left(e^{it}\right) du\left(t\right) + f\left(e^{\frac{x+b}{2}i}\right) \int_x^b g\left(e^{it}\right) du\left(t\right)$$

$$+ \int_a^x \left[f\left(e^{it}\right) - f\left(e^{\frac{a+x}{2}i}\right)\right] g\left(e^{it}\right) du\left(t\right)$$

$$+ \int_x^b \left[f\left(e^{it}\right) - f\left(e^{\frac{x+b}{2}i}\right)\right] g\left(e^{it}\right) du\left(t\right). \quad (7.12)$$

Also, if we take $\lambda = \frac{f\left(e^{ia}\right) + f\left(e^{ix}\right)}{2}$ and $\mu = \frac{f\left(e^{ix}\right) + f\left(e^{ib}\right)}{2}$ in (7.1) we get for $x \in [a, b]$ that

$$\int_a^b f\left(e^{it}\right) g\left(e^{it}\right) du\left(t\right)$$

$$= \frac{f\left(e^{ia}\right) + f\left(e^{ix}\right)}{2} \int_a^x g\left(e^{it}\right) du\left(t\right)$$

$$+ \frac{f\left(e^{ix}\right) + f\left(e^{ib}\right)}{2} \int_x^b g\left(e^{it}\right) du\left(t\right)$$

$$+ \int_a^x \left[f\left(e^{it}\right) - \frac{f\left(e^{ia}\right) + f\left(e^{ix}\right)}{2}\right] g\left(e^{it}\right) du\left(t\right)$$

$$+ \int_x^b \left[f\left(e^{it}\right) - \frac{f\left(e^{ix}\right) + f\left(e^{ib}\right)}{2}\right] g\left(e^{it}\right) du\left(t\right) \quad (7.13)$$

while for $\lambda = \frac{1}{x-a} \int_a^x f\left(e^{is}\right) ds$ and $\mu = \frac{1}{b-x} \int_x^b f\left(e^{is}\right) ds$ we get

$$
\int_a^b f\left(e^{it}\right) g\left(e^{it}\right) du(t)
$$
$$
= \frac{1}{x-a} \int_a^x f\left(e^{is}\right) ds \int_a^x g\left(e^{it}\right) du(t)
$$
$$
+ \frac{1}{b-x} \int_x^b f\left(e^{is}\right) ds \int_x^b g\left(e^{it}\right) du(t)
$$
$$
+ \int_a^x \left[f\left(e^{it}\right) - \frac{1}{x-a} \int_a^x f\left(e^{is}\right) ds \right] g\left(e^{it}\right) du(t)
$$
$$
+ \int_x^b \left[f\left(e^{it}\right) - \frac{1}{b-x} \int_x^b f\left(e^{is}\right) ds \right] g\left(e^{it}\right) du(t) \quad (7.14)
$$

for any $x \in (a, b)$.

In particular, for $g\left(e^{it}\right) = 1$, $t \in [a, b]$, we have for $x \in [a, b]$ that

$$
\int_a^b f\left(e^{it}\right) du(t) = [u(x) - u(a)] f\left(e^{ia}\right) + [u(b) - u(x)] f\left(e^{ib}\right)
$$
$$
+ \int_a^x \left[f\left(e^{it}\right) - f\left(e^{ia}\right) \right] du(t) + \int_x^b \left[f\left(e^{it}\right) - f\left(e^{ib}\right) \right] du(t),
$$
$$
(7.15)
$$

$$
\int_a^b f\left(e^{it}\right) du(t)
$$
$$
= [u(x) - u(a)] f\left(e^{\frac{a+x}{2}i}\right) + [u(b) - u(x)] f\left(e^{\frac{x+b}{2}i}\right)
$$
$$
+ \int_a^x \left[f\left(e^{it}\right) - f\left(e^{\frac{a+x}{2}i}\right) \right] du(t)
$$
$$
+ \int_x^b \left[f\left(e^{it}\right) - f\left(e^{\frac{x+b}{2}i}\right) \right] du(t), \quad (7.16)
$$

$$\int_a^b f\left(e^{it}\right) du\left(t\right)$$

$$= \left[u\left(x\right) - u\left(a\right)\right] \frac{f\left(e^{ia}\right) + f\left(e^{ix}\right)}{2}$$

$$+ \left[u\left(b\right) - u\left(x\right)\right] \frac{f\left(e^{ix}\right) + f\left(e^{ib}\right)}{2}$$

$$+ \int_a^x \left[f\left(e^{it}\right) - \frac{f\left(e^{ia}\right) + f\left(e^{ix}\right)}{2}\right] du\left(t\right)$$

$$+ \int_x^b \left[f\left(e^{it}\right) - \frac{f\left(e^{ix}\right) + f\left(e^{ib}\right)}{2}\right] du\left(t\right) \quad (7.17)$$

and that

$$\int_a^b f\left(e^{it}\right) du\left(t\right)$$

$$= \frac{u\left(x\right) - u\left(a\right)}{x - a} \int_a^x f\left(e^{is}\right) ds + \frac{u\left(b\right) - u\left(x\right)}{b - x} \int_x^b f\left(e^{is}\right) ds$$

$$+ \int_a^x \left[f\left(e^{it}\right) - \frac{1}{x - a} \int_a^x f\left(e^{is}\right) ds\right] du\left(t\right)$$

$$+ \int_x^b \left[f\left(e^{it}\right) - \frac{1}{b - x} \int_x^b f\left(e^{is}\right) ds\right] du\left(t\right). \quad (7.18)$$

7.2 INEQUALITIES FOR BOUNDED FUNCTIONS

Now, for γ, $\Gamma \in \mathbb{C}$, and $[a, b] \subset [0, 2\pi]$, an interval of real numbers, define the sets of complex-valued functions

$$\bar{U}_{\exp[a,b]}\left(\gamma, \Gamma\right)$$

$$:= \left\{f : \mathcal{C}\left(0, 1\right) \to \mathbb{C} \mid \operatorname{Re}\left[\left(\Gamma - f\left(e^{it}\right)\right) \left(\overline{f\left(e^{it}\right)} - \overline{\gamma}\right)\right] \geq 0\right.$$

$$\left.\text{for each } t \in [a, b]\right\}$$

and

$$\bar{\Delta}_{\exp[a,b]}(\gamma, \Gamma)$$

$$:= \left\{ f : \mathcal{C}(0,1) \to \mathbb{C} \mid \left| f\left(e^{it}\right) - \frac{\gamma + \Gamma}{2} \right| \right.$$

$$\left. \leq \frac{1}{2} |\Gamma - \gamma| \text{ for each } t \in [a,b] \right\}.$$

The following representation result may be stated.

Proposition 7.3 *For any* γ, $\Gamma \in \mathbb{C}$, $\gamma \neq \Gamma$, *we have that* $\bar{U}_{\exp[a,b]}(\gamma, \Gamma)$ *and* $\bar{\Delta}_{\exp[a,b]}(\gamma, \Gamma)$ *are nonempty, convex and closed sets and*

$$\bar{U}_{\exp[a,b]}(\gamma, \Gamma) = \bar{\Delta}_{\exp[a,b]}(\gamma, \Gamma). \tag{7.19}$$

Proof. We observe that for any $z \in \mathbb{C}$ we have the equivalence

$$\left| z - \frac{\gamma + \Gamma}{2} \right| \leq \frac{1}{2} |\Gamma - \gamma|$$

if and only if

$$\text{Re} \left[(\Gamma - z)(\bar{z} - \bar{\gamma}) \right] \geq 0.$$

This follows by the equality

$$\frac{1}{4} |\Gamma - \gamma|^2 - \left| z - \frac{\gamma + \Gamma}{2} \right|^2 = \text{Re} \left[(\Gamma - z)(\bar{z} - \bar{\gamma}) \right]$$

that holds for any $z \in \mathbb{C}$.

The equality (7.19) is thus a simple consequence of this fact.

On making use of the complex numbers field properties we can also state that:

Corollary 7.4 *For any* γ, $\Gamma \in \mathbb{C}$, $\gamma \neq \Gamma$, *we have that*

$$\bar{U}_{\exp[a,b]}(\gamma, \Gamma)$$

$$= \left\{ f : \mathcal{C}(0,1) \to \mathbb{C} \mid \left(\text{Re}\,\Gamma - \text{Re}\,f\left(e^{it}\right) \right) \left(\text{Re}\,f\left(e^{it}\right) - \text{Re}\,\gamma \right) \right.$$

$$+ \left(\text{Im}\,\Gamma - \text{Im}\,f\left(e^{it}\right) \right) \left(\text{Im}\,f\left(e^{it}\right) - \text{Im}\,\gamma \right) \geq 0$$

$$\left. \text{for each } t \in [a,b] \right\}. \tag{7.20}$$

Now, if we assume that $\mathrm{Re}(\Gamma) \geq \mathrm{Re}(\gamma)$ and $\mathrm{Im}(\Gamma) \geq \mathrm{Im}(\gamma)$, then we can define the following set of functions as well:

$$\bar{S}_{\exp[a,b]}(\gamma, \Gamma)$$
$$:= \left\{ f : \mathcal{C}(0,1) \to \mathbb{C} \mid \mathrm{Re}(\Gamma) \geq \mathrm{Re}\, f\left(e^{it}\right) \geq \mathrm{Re}(\gamma) \right.$$
$$\left. \text{and } \mathrm{Im}(\Gamma) \geq \mathrm{Im}\, f\left(e^{it}\right) \geq \mathrm{Im}(\gamma) \text{ for each } t \in [a,b] \right\}.$$
$$(7.21)$$

One can easily observe that $\bar{S}_{\exp[a,b]}(\gamma, \Gamma)$ is closed, convex and

$$\emptyset \neq \bar{S}_{\exp[a,b]}(\gamma, \Gamma) \subseteq \bar{U}_{\exp[a,b]}(\gamma, \Gamma). \qquad (7.22)$$

We consider the following functional

$$P_{\exp}(f, g, v; \gamma, \Gamma, a, b)$$
$$:= \int_a^b f\left(e^{it}\right) g\left(e^{it}\right) dv(t) - \frac{\gamma + \Gamma}{2} \int_a^b g\left(e^{it}\right) dv(t) \quad (7.23)$$

for the complex-valued functions f, $g \in \mathcal{C}(0,1)$, v defined on $[a,b]$ and such that the involved Riemann–Stieltjes integrals exist, and for γ, $\Gamma \in \mathbb{C}$.

Theorem 7.5 (Dragomir 2019, [23]) *Let f, $g \in \mathcal{C}(0,1)$ and γ, $\Gamma \in \mathbb{C}$, $\gamma \neq \Gamma$ such that $f \in \bar{\Delta}_{\exp[a,b]}(\gamma, \Gamma)$.*

(i) If $v \in \mathcal{BV}_{\mathbb{C}}[a,b]$, then

$$|P_{\exp}(f, g, v; \gamma, \Gamma, a, b)|$$
$$\leq \frac{1}{2}|\Gamma - \gamma| \int_a^b \left|g\left(e^{it}\right)\right| d\left(\bigvee_a^t (v)\right)$$
$$\leq \frac{1}{2}|\Gamma - \gamma| \max_{t \in [a,b]} \left|g\left(e^{it}\right)\right| \bigvee_a^b (v). \qquad (7.24)$$

(ii) If $v \in \mathcal{L}_{L,\mathbb{C}}[a,b]$, namely, v is Lipschitzian with the constant $L > 0$,

$$|v(t) - v(s)| \leq L|t - s| \text{ for any } t, \; s \in [a,b],$$

then we also have

$$|P_{\exp}(f, g, v; \gamma, \Gamma, a, b)| \leq \frac{1}{2} |\Gamma - \gamma| L \int_a^b \left| g\left(e^{it}\right) \right| dt$$

$$\leq \frac{1}{2} |\Gamma - \gamma| (b - a) \max_{t \in [a,b]} \left| g\left(e^{it}\right) \right|. \quad (7.25)$$

(iii) If $v \in \mathcal{M}^{\nearrow}[a, b]$, namely, v is monotonic increasing on $[a, b]$, then we have

$$|P_{\exp}(f, g, v; \gamma, \Gamma, a, b)| \leq \frac{1}{2} |\Gamma - \gamma| \int_a^b \left| g\left(e^{it}\right) \right| dv(t)$$

$$\leq \frac{1}{2} |\Gamma - \gamma| [v(b) - v(a)] \max_{t \in [a,b]} \left| g\left(e^{it}\right) \right|. \quad (7.26)$$

Proof. (i) It is well known that if $p \in \mathcal{R}(u, [a, b])$ where $u \in \mathcal{BV}_{\mathbb{C}}[a, b]$ then we have [1, p. 177]

$$\left| \int_a^b p(t) \, du(t) \right| \leq \int_a^b |p(t)| \, d\left(\bigvee_a^t (u) \right) \leq \sup_{t \in [a,b]} |p(t)| \bigvee_a^b (u). \quad (7.27)$$

By the equality (7.2) we have

$$\int_a^b f\left(e^{it}\right) g\left(e^{it}\right) dv(t) - \frac{\gamma + \Gamma}{2} \int_a^b g\left(e^{it}\right) dv(t)$$

$$= \int_a^b \left[f\left(e^{it}\right) - \frac{\gamma + \Gamma}{2} \right] g\left(e^{it}\right) dv(t). \quad (7.28)$$

Since $f \in \bar{\Delta}_{\exp[a,b]}(\gamma, \Gamma)$ then by (7.27) and (7.28) we have

$$\left| \int_a^b f\left(e^{it}\right) g\left(e^{it}\right) dv(t) - \frac{\gamma + \Gamma}{2} \int_a^b g\left(e^{it}\right) dv(t) \right|$$

$$\leq \int_a^b \left| \left[f\left(e^{it}\right) - \frac{\gamma + \Gamma}{2} \right] g\left(e^{it}\right) \right| d\left(\bigvee_a^t (v) \right)$$

$$= \int_a^b \left| f\left(e^{it}\right) - \frac{\gamma + \Gamma}{2} \right| \left| g\left(e^{it}\right) \right| d\left(\bigvee_a^t (v) \right)$$

$$\leq \frac{1}{2} |\Gamma - \gamma| \int_a^b \left| g\left(e^{it}\right) \right| d\left(\bigvee_a^t (v) \right)$$

and the first inequality in (7.24) is proved. The second part is obvious.

(ii) It is well known that if $p \in \mathcal{R}(v, [a, b])$, where $v \in \mathcal{L}_{L, \mathbb{C}}[a, b]$, then we have

$$\left| \int_a^b p(t) \, dv(t) \right| \leq L \int_a^b |p(t)| \, dt. \qquad (7.29)$$

By using (7.28) we then get (7.25).

(iii) It is well known that if $p \in \mathcal{R}(v, [a, b])$, where $v \in \mathcal{M}^{\nearrow}[a, b]$, then we have

$$\left| \int_a^b p(t) \, dv(t) \right| \leq \int_a^b |p(t)| \, dv(t). \qquad (7.30)$$

By using (7.29) we then get (7.26).

Remark 30 *We define the simpler functional for $g \equiv 1$ by*

$$P_{\exp}(f, v; \gamma, \Gamma, a, b) := P_{\exp}(f, 1, v; \gamma, \Gamma, a, b)$$
$$= \int_a^b f\left(e^{it}\right) dv(t) - \frac{\gamma + \Gamma}{2} \left[v(b) - v(a)\right].$$

Let $f \in \mathcal{C}(0, 1)$ and $\gamma, \Gamma \in \mathbb{C}$, $\gamma \neq \Gamma$ such that $f \in \bar{\Delta}_{\exp[a,b]}(\gamma, \Gamma)$. If $v \in \mathcal{BV}_{\mathbb{C}}[a, b]$, then

$$|P_{\exp}(f, v; \gamma, \Gamma, a, b)| \leq \frac{1}{2} |\Gamma - \gamma| \bigvee_a^b (v). \qquad (7.31)$$

If $v \in \mathcal{L}_{L, \mathbb{C}}[a, b]$, then

$$|P_{\exp}(f, v; \gamma, \Gamma, a, b)| \leq \frac{1}{2} L |\Gamma - \gamma| (b - a). \qquad (7.32)$$

If $v \in \mathcal{M}^{\nearrow}[a, b]$, then

$$|P_{\exp}(f, v; \gamma, \Gamma, a, b)| \leq \frac{1}{2} |\Gamma - \gamma| [v(b) - v(a)]. \qquad (7.33)$$

7.3 QUASI-GRÜSS-TYPE INEQUALITIES

We consider the functional

$$Q_{\exp}(f, g, v; \gamma, \Gamma, \delta, \Delta, a, b)$$
$$:= \int_a^b f\left(e^{it}\right) g\left(e^{it}\right) dv(t) - \frac{\gamma + \Gamma}{2} \int_a^b g\left(e^{it}\right) dv(t)$$
$$- \frac{\delta + \Delta}{2} \int_a^b f\left(e^{it}\right) dv(t) + \frac{\gamma + \Gamma}{2} \cdot \frac{\delta + \Delta}{2} [v(b) - v(a)]$$

$$(7.34)$$

for the complex-valued functions f, $g \in \mathcal{C}(0,1)$ and v defined on $[a, b] \subset [0, 2\pi]$ and such that the involved Riemann–Stieltjes integrals exist, and for γ, Γ, δ, $\Delta \in \mathbb{C}$.

We have the following quasi-Grüss-type inequality:

Proposition 7.6 (Dragomir 2019, [23]) *Let $f, g \in \mathcal{C}(0,1)$ and γ, Γ, δ, $\Delta \in \mathbb{C}$, $\gamma \neq \Gamma$, $\delta \neq \Delta$ such that $f \in \bar{\Delta}_{\exp[a,b]}(\gamma, \Gamma)$ and $g \in \bar{\Delta}_{\exp[a,b]}(\delta, \Delta)$. If $v \in \mathcal{BV}_{\mathbb{C}}[a, b]$, then*

$$|Q_{\exp}(f, g, v; \gamma, \Gamma, a, b)| \leq \frac{1}{4} |\Gamma - \gamma| |\Delta - \delta| \bigvee_a^b (v). \quad (7.35)$$

If $v \in \mathcal{L}_{L,\mathbb{C}}[a, b]$, then

$$|Q_{\exp}(f, g, v; \gamma, \Gamma, a, b)| \leq \frac{1}{4} |\Gamma - \gamma| |\Delta - \delta| L(b - a).$$

If $v \in \mathcal{M}^{\nearrow}[a, b]$, then

$$|Q_{\exp}(f, g, v; \gamma, \Gamma, a, b)| \leq \frac{1}{4} |\Gamma - \gamma| |\Delta - \delta| [v(b) - v(a)].$$

Proof. If we replace in (7.12) $g\left(e^{i\cdot}\right)$ by $g\left(e^{i\cdot}\right) - \frac{\delta + \Delta}{2}$, then we get

$$\left| \int_a^b f\left(e^{it}\right) g\left(e^{it}\right) dv(t) - \frac{\gamma + \Gamma}{2} \int_a^b g\left(e^{it}\right) dv(t) \right.$$

$$\left. - \frac{\delta + \Delta}{2} \int_a^b f\left(e^{it}\right) dv(t) + \frac{\gamma + \Gamma}{2} \cdot \frac{\delta + \Delta}{2} [v(b) - v(a)] \right|$$

$$\leq \frac{1}{2} |\Gamma - \gamma| \int_a^b \left| g\left(e^{it}\right) - \frac{\delta + \Delta}{2} \right| d\left(\bigvee_a^t (v) \right).$$

Since $g \in \bar{\Delta}_{\exp[a,b]}(\delta, \Delta)$, then

$$\int_a^b \left| g\left(e^{it}\right) - \frac{\delta + \Delta}{2} \right| d\left(\bigvee_a^t (v) \right) \leq \frac{1}{2} |\Delta - \delta| \int_a^b d\left(\bigvee_a^t (v) \right)$$

$$= \frac{1}{2} |\Delta - \delta| \bigvee_a^b (v)$$

and the inequality (7.35) is proved.

The proofs of the other two statements follow in a similar way and we omit the details.

Proposition 7.7 (Dragomir 2019, [23]) *Let* $f, g \in \mathcal{C}$ $(0,1)$, $g\left(e^{i\cdot}\right) \in \mathcal{BV}_{\mathbb{C}}[a,b]$ *and* $\gamma, \Gamma \in \mathbb{C}$, $\gamma \neq \Gamma$ *such that* $f \in \bar{\Delta}_{\exp[a,b]}(\gamma, \Gamma)$. *If* $v \in \mathcal{BV}_{\mathbb{C}}[a,b]$, *then*

$$\left| Q_{\exp}\left(f, g, v; g\left(e^{ia}\right), g\left(e^{ib}\right), a, b\right) \right|$$

$$\leq \frac{1}{4} |\Gamma - \gamma| \bigvee_a^b \left(g\left(e^{i\cdot}\right)\right) \bigvee_a^b (v), \qquad (7.36)$$

where

$$Q_{\exp}\left(f, g, v; g\left(e^{ia}\right), g\left(e^{ib}\right), a, b\right)$$

$$= \int_a^b f\left(e^{it}\right) g\left(e^{it}\right) dv(t) - \frac{\gamma + \Gamma}{2} \int_a^b g\left(e^{it}\right) dv(t)$$

$$- \frac{g\left(e^{ia}\right) + g\left(e^{ib}\right)}{2} \int_a^b f\left(e^{it}\right) dv(t)$$

$$+ \frac{\gamma + \Gamma}{2} \cdot \frac{g\left(e^{ia}\right) + g\left(e^{ib}\right)}{2} [v(b) - v(a)].$$

If $v \in \mathcal{L}_{L,\mathbb{C}}[a,b]$, then

$$\left| Q_{\exp}\left(f,g,v;g\left(e^{ia}\right),g\left(e^{ib}\right),a,b\right)\right|$$

$$\leq \frac{1}{4}\left|\Gamma-\gamma\right|L\left(b-a\right)\bigvee_a^b\left(g\left(e^{i\cdot}\right)\right). \quad (7.37)$$

If $v \in \mathcal{M}^{\nearrow}[a,b]$, then

$$\left| Q_{\exp}\left(f,g,v;g\left(e^{ia}\right),g\left(e^{ib}\right),a,b\right)\right|$$

$$\leq \frac{1}{4}\left|\Gamma-\gamma\right|\bigvee_a^b\left(g\left(e^{i\cdot}\right)\right)\left[v\left(b\right)-v\left(a\right)\right]. \quad (7.38)$$

Proof. If we replace in (7.12) $g\left(e^{i\cdot}\right)$ by $g\left(e^{i\cdot}\right) - \frac{g\left(e^{ia}\right)+g\left(e^{ib}\right)}{2}$, then we get

$$\left|\int_a^b f\left(e^{it}\right)g\left(e^{it}\right)dv\left(t\right) - \frac{\gamma+\Gamma}{2}\int_a^b g\left(e^{it}\right)dv\left(t\right)\right.$$

$$-\frac{g\left(e^{ia}\right)+g\left(e^{ib}\right)}{2}\int_a^b f\left(e^{it}\right)dv\left(t\right)$$

$$\left.+\frac{\gamma+\Gamma}{2}\cdot\frac{g\left(e^{ia}\right)+g\left(e^{ib}\right)}{2}\left[v\left(b\right)-v\left(a\right)\right]\right|$$

$$\leq \frac{1}{2}\left|\Gamma-\gamma\right|\int_a^b\left|g\left(e^{it}\right)-\frac{g\left(e^{ia}\right)+g\left(e^{ib}\right)}{2}\right|d\left(\bigvee_a^t\left(v\right)\right).$$

Since $g\left(e^{i\cdot}\right) \in \mathcal{BV}_{\mathbb{C}}[a,b]$, hence

$$\left|g\left(e^{it}\right)-\frac{g\left(e^{ia}\right)+g\left(e^{ib}\right)}{2}\right|$$

$$=\left|\frac{g\left(e^{it}\right)-g\left(e^{ia}\right)+g\left(e^{it}\right)-g\left(e^{ib}\right)}{2}\right|$$

$$\leq \frac{1}{2}\left[\left|g\left(e^{it}\right)-g\left(e^{ia}\right)\right|+\left|g\left(e^{ib}\right)-g\left(e^{it}\right)\right|\right]\leq\frac{1}{2}\bigvee_a^b\left(g\left(e^{i\cdot}\right)\right)$$

for any $t \in [a, b]$.

Therefore

$$\int_a^b \left| g\left(e^{it}\right) - \frac{g\left(e^{ia}\right) + g\left(e^{ib}\right)}{2} \right| d\left(\bigvee_a^t (v)\right)$$

$$\leq \frac{1}{2} \bigvee_a^b \left(g\left(e^{i\cdot}\right)\right) \int_a^b d\left(\bigvee_a^t (v)\right) = \frac{1}{2} \bigvee_a^b \left(g\left(e^{i\cdot}\right)\right) \bigvee_a^b (v)$$

and the inequality (7.36) is proved.

The proofs of the other statements follow in a similar way and we omit the details.

Proposition 7.8 (Dragomir 2019, [23]) *Let* $f, g \in \mathcal{C}(0, 1)$ *and* $\gamma, \Gamma \in \mathbb{C}, \gamma \neq \Gamma$ *such that* $f \in \bar{\Delta}_{\exp[a,b]}(\gamma, \Gamma)$. *If* $v \in \mathcal{BV}_{\mathbb{C}}[a, b]$, *then*

$$\left| Q_{\exp}\left(f, g, v; \frac{1}{b-a}\int_a^b g\left(e^{is}\right) ds, \frac{1}{b-a}\int_a^b g\left(e^{is}\right) ds, a, b\right) \right|$$

$$\leq \frac{1}{2} |\Gamma - \gamma| \int_a^b \left| g\left(e^{it}\right) - \frac{1}{b-a}\int_a^b g\left(e^{is}\right) ds \right| d\left(\bigvee_a^t (v)\right)$$

$$\leq \frac{1}{2} |\Gamma - \gamma| \max_{t \in [a,b]} \left| g\left(e^{it}\right) - \frac{1}{b-a}\int_a^b g\left(e^{is}\right) ds \right| \bigvee_a^b (v),$$

$$(7.39)$$

where

$$Q_{\exp}\left(f, g, v; \frac{1}{b-a}\int_a^b g\left(e^{is}\right) ds, \frac{1}{b-a}\int_a^b g\left(e^{is}\right) ds, a, b\right)$$

$$= \int_a^b f\left(e^{it}\right) g\left(e^{it}\right) dv(t) - \frac{\gamma + \Gamma}{2} \int_a^b g\left(e^{it}\right) dv(t)$$

$$- \int_a^b f\left(e^{it}\right) dv(t) \frac{1}{b-a}\int_a^b g\left(e^{it}\right) dt$$

$$+ [v(b) - v(a)] \frac{\gamma + \Gamma}{2} \cdot \frac{1}{b-a}\int_a^b g\left(e^{it}\right) dt.$$

If $v \in \mathcal{L}_{L,\mathbb{C}}[a,b]$, then

$$\left| Q_{\exp}\left(f, g, v; \frac{1}{b-a}\int_a^b g\left(e^{is}\right)ds, \frac{1}{b-a}\int_a^b g\left(e^{is}\right)ds, a, b\right)\right|$$

$$\leq \frac{1}{2}\left|\Gamma - \gamma\right| L \int_a^b \left| g\left(e^{it}\right) - \frac{1}{b-a}\int_a^b g\left(e^{is}\right)ds\right| dt$$

$$\leq \frac{1}{2}\left|\Gamma - \gamma\right| L (b-a) \max_{t\in[a,b]}\left| g\left(e^{it}\right) - \frac{1}{b-a}\int_a^b g\left(e^{is}\right)ds\right| dt.$$

$$(7.40)$$

If $v \in \mathcal{M}^{\nearrow}[a,b]$, then

$$\left| Q_{\exp}\left(f, g, v; \frac{1}{b-a}\int_a^b g\left(e^{is}\right)ds, \frac{1}{b-a}\int_a^b g\left(e^{is}\right)ds, a, b\right)\right|$$

$$\leq \frac{1}{2}\left|\Gamma - \gamma\right| \int_a^b \left| g\left(e^{it}\right) - \frac{1}{b-a}\int_a^b g\left(e^{is}\right)ds\right| dv(t)$$

$$\leq \frac{1}{2}\left|\Gamma - \gamma\right| \max_{t\in[a,b]}\left| g\left(e^{it}\right) - \frac{1}{b-a}\int_a^b g\left(e^{is}\right)ds\right| [v(b) - v(a)].$$

$$(7.41)$$

Proof. The first inequality follows by Theorem 7.5 by replacing $g\left(e^{i\cdot}\right)$ with $g\left(e^{i\cdot}\right) - \frac{1}{b-a}\int_a^b g\left(e^{is}\right)ds$. The second part follows by the fact that

$$\int_a^b \left| g\left(e^{it}\right) - \frac{1}{b-a}\int_a^b g\left(e^{is}\right)ds\right| d\left(\bigvee_a^t (v)\right)$$

$$\leq \max_{t\in[a,b]}\left| g\left(e^{it}\right) - \frac{1}{b-a}\int_a^b g\left(e^{is}\right)ds\right| \int_a^b d\left(\bigvee_a^t (v)\right)$$

$$= \max_{t\in[a,b]}\left| g\left(e^{it}\right) - \frac{1}{b-a}\int_a^b g\left(e^{is}\right)ds\right| \bigvee_a^b (v).$$

The proofs of the other statements follow in a similar way and we omit the details.

Remark 31 *We observe that the quantity*

$$\left| g\left(e^{it} \right) - \frac{1}{b-a} \int_a^b g\left(e^{is} \right) ds \right| , \ t \in [a,b]$$

is the left-hand side in Ostrowski-type inequalities for various classes of functions g. For a recent survey on these inequalities, see [22]. Therefore, if

$$\left| g\left(e^{it} \right) - \frac{1}{b-a} \int_a^b g\left(e^{is} \right) ds \right| \le M_{g,[a,b]}(t), \ t \in [a,b]$$

is such an inequality, then from (7.39) we get

$$\left| Q_{\exp}\left(f,g,v; \frac{1}{b-a} \int_a^b g\left(e^{is} \right) ds, \frac{1}{b-a} \int_a^b g\left(e^{is} \right) ds, a, b \right) \right|$$

$$\le \frac{1}{2} |\Gamma - \gamma| \int_a^b M_{g,[a,b]}(t) \, d\left(\bigvee_a^t (v) \right) \quad (7.42)$$

if $v \in \mathcal{BV}_{\mathbb{C}}[a,b]$, from (7.40) we get

$$\left| Q_{\exp}\left(f,g,v; \frac{1}{b-a} \int_a^b g\left(e^{is} \right) ds, \frac{1}{b-a} \int_a^b g\left(e^{is} \right) ds, a, b \right) \right|$$

$$\le \frac{1}{2} |\Gamma - \gamma| L \int_a^b M_{g,[a,b]}(t) \, dt \quad (7.43)$$

if $v \in \mathcal{L}_{L,\mathbb{C}}[a,b]$ and from (7.41) we get

$$\left| Q_{\exp}\left(f,g,v; \frac{1}{b-a} \int_a^b g\left(e^{is} \right) ds, \frac{1}{b-a} \int_a^b g\left(e^{is} \right) ds, a, b \right) \right|$$

$$\le \frac{1}{2} |\Gamma - \gamma| \int_a^b M_{g,[a,b]}(t) \, dv(t), \quad (7.44)$$

if $v \in \mathcal{M}^{\nearrow}[a,b]$.

For instance, if $g\left(e^{i \cdot} \right) : [a,b] \to \mathbb{C}$ is of bounded variation, then we have, see [7] and [10],

$$\left| g\left(e^{it} \right) - \frac{1}{b-a} \int_a^b g\left(e^{is} \right) ds \right| \le \left[\frac{1}{2} + \frac{\left| t - \frac{a+b}{2} \right|}{b-a} \right] \bigvee_a^b \left(g\left(e^{i \cdot} \right) \right)$$

$$(7.45)$$

for any $t \in [a, b]$. The constant $\frac{1}{2}$ is the best possible one. Observe that

$$\int_a^b \left[\frac{1}{2} + \frac{\left| t - \frac{a+b}{2} \right|}{b-a} \right] d \left(\bigvee_a^t (v) \right)$$

$$= \frac{1}{2} \bigvee_a^b (v) + \frac{1}{b-a} \int_a^b \left| t - \frac{a+b}{2} \right| d \left(\bigvee_a^t (v) \right)$$

$$= \frac{1}{2} \bigvee_a^b (v) + \frac{1}{b-a} \int_a^{\frac{a+b}{2}} \left(\frac{a+b}{2} - t \right) d \left(\bigvee_a^t (v) \right)$$

$$+ \frac{1}{b-a} \int_{\frac{a+b}{2}}^b \left(t - \frac{a+b}{2} \right) d \left(\bigvee_a^t (v) \right)$$

$$= \frac{1}{2} \bigvee_a^b (v) + \frac{1}{b-a} \left[\left(\frac{a+b}{2} - t \right) \bigvee_a^t (v) \Big|_a^{\frac{a+b}{2}} + \int_a^{\frac{a+b}{2}} \bigvee_a^t (v) \, dt \right]$$

$$+ \frac{1}{b-a} \left[\left(t - \frac{a+b}{2} \right) \bigvee_a^t (v) \Big|_{\frac{a+b}{2}}^b - \int_{\frac{a+b}{2}}^b \bigvee_a^t (v) \, dt \right]$$

$$= \bigvee_a^b (v) + \frac{1}{b-a} \int_a^{\frac{a+b}{2}} \bigvee_a^t (v) \, dt - \frac{1}{b-a} \int_{\frac{a+b}{2}}^b \bigvee_a^t (v) \, dt$$

$$= \bigvee_a^b (v) + \frac{1}{b-a} \left(\int_a^{\frac{a+b}{2}} \bigvee_a^t (v) \, dt - \int_{\frac{a+b}{2}}^b \bigvee_a^t (v) \, dt \right)$$

$$= \bigvee_a^b (v) - \frac{1}{b-a} \int_a^b \operatorname{sgn} \left(t - \frac{a+b}{2} \right) \bigvee_a^t (v) \, dt.$$

Then by (2.32) we get

$$\left| Q_{\exp}\left(f, g, v; \frac{1}{b-a} \int_a^b g\left(e^{is}\right) ds, \frac{1}{b-a} \int_a^b g\left(e^{is}\right) ds, a, b \right) \right|$$

$$\leq \frac{1}{2} |\Gamma - \gamma| \bigvee_a^b \left(g\left(e^{i\cdot}\right) \right)$$

$$\left[\bigvee_a^b (v) - \frac{1}{b-a} \int_a^b \operatorname{sgn}\left(t - \frac{a+b}{2} \right) \bigvee_a^t (v) dt \right]$$

$$\leq \frac{1}{2} |\Gamma - \gamma| \bigvee_a^b \left(g\left(e^{i\cdot}\right) \right) \bigvee_a^b (v) \quad (7.46)$$

if $v, g\left(e^{i\cdot}\right) \in \mathcal{BV}_{\mathbb{C}}[a, b]$.

The last inequality in (7.46) follows by Chebyshev's inequality for monotonic functions that gives that

$$\frac{1}{b-a} \int_a^b \operatorname{sgn}\left(t - \frac{a+b}{2} \right) \bigvee_a^t (v) dt$$

$$\geq \frac{1}{b-a} \int_a^b \operatorname{sgn}\left(t - \frac{a+b}{2} \right) dt \frac{1}{b-a} \int_a^b \bigvee_a^t (v) dt = 0.$$

Observe also that

$$\int_a^b \left[\frac{1}{2} + \frac{\left| t - \frac{a+b}{2} \right|}{b-a} \right] dt = \frac{3}{4} (b-a),$$

then by (2.33) we get

$$\left| Q_{\exp}\left(f, g, v; \frac{1}{b-a} \int_a^b g\left(e^{is}\right) ds, \frac{1}{b-a} \int_a^b g\left(e^{is}\right) ds, a, b \right) \right|$$

$$\leq \frac{3}{8} |\Gamma - \gamma| L (b-a) \bigvee_a^b \left(g\left(e^{i\cdot}\right) \right) \quad (7.47)$$

if $v \in \mathcal{L}_{L,\mathbb{C}}[a, b]$ and $g\left(e^{i\cdot}\right) \in \mathcal{BV}_{\mathbb{C}}[a, b]$.

Finally, since

$$\int_a^b \left[\frac{1}{2} + \frac{\left| t - \frac{a+b}{2} \right|}{b-a} \right] dv(t) = v(b) - v(a)$$

$$- \frac{1}{b-a} \int_a^b \mathrm{sgn}\left(t - \frac{a+b}{2} \right) [v(t) - v(a)]\, dt,$$

then we get by (2.34) that

$$\left| Q_{\exp}\left(f, g, v; \frac{1}{b-a} \int_a^b g\left(e^{is}\right) ds, \frac{1}{b-a} \int_a^b g\left(e^{is}\right) ds, a, b \right) \right|$$

$$\leq \frac{1}{2} |\Gamma - \gamma| \bigvee_a^b \left(g\left(e^{i \cdot}\right) \right)$$

$$\times \left[v(b) - v(a) - \frac{1}{b-a} \int_a^b \mathrm{sgn}\left(t - \frac{a+b}{2} \right) [v(t) - v(a)]\, dt \right]$$

$$\leq \frac{1}{2} |\Gamma - \gamma| \bigvee_a^b \left(g\left(e^{i \cdot}\right) \right) [v(b) - v(a)] \quad (7.48)$$

if $v \in \mathcal{M}^{\nearrow}[a,b]$ and $g\left(e^{i \cdot}\right) \in \mathcal{BV}_{\mathbb{C}}[a,b]$.

7.4 GRÜSS-TYPE INEQUALITIES

Consider the *Grüss-type functional*

$$G_{\exp}(f, g, v; a, b) := \int_a^b f\left(e^{it}\right) g\left(e^{it}\right) dv(t)$$

$$- \frac{1}{v(b) - v(a)} \int_a^b f\left(e^{it}\right) dv(t) \int_a^b g\left(e^{it}\right) dv(t) \quad (7.49)$$

for the complex-valued functions $f, g \in \mathcal{C}(0,1)$, v defined on $[a,b]$ and such that the involved Riemann–Stieltjes integrals exist and $v(b) \neq v(a)$.

We have:

Proposition 7.9 (Dragomir 2019, [23]) *Let $f, g \in \mathcal{C}(0,1)$*

and $\gamma, \Gamma \in \mathbb{C}$, $\gamma \neq \Gamma$ *such that* $f \in \bar{\Delta}_{\exp[a,b]}(\gamma, \Gamma)$. *If* $v \in \mathcal{BV}_{\mathbb{C}}[a,b]$ *with* $v(b) \neq v(a)$, *then*

$$|G_{\exp}(f,g,v;a,b)|$$

$$\leq \frac{1}{2}|\Gamma - \gamma| \int_a^b \left| g\left(e^{it}\right) - \frac{1}{v(b) - v(a)} \int_a^b g\left(e^{is}\right) dv(s) \right|$$

$$d\left(\bigvee_a^t (v)\right) \leq \frac{1}{2}|\Gamma - \gamma| \bigvee_a^b (v) \max_{t \in [a,b]}$$

$$\left| g\left(e^{it}\right) - \frac{1}{v(b) - v(a)} \int_a^b g\left(e^{is}\right) dv(s) \right|. \quad (7.50)$$

If $v \in \mathcal{L}_{L,\mathbb{C}}[a,b]$, *then*

$$|G_{\exp}(f,g,v;a,b)|$$

$$\leq \frac{1}{2}|\Gamma - \gamma| L \int_a^b \left| g\left(e^{it}\right) - \frac{1}{v(b) - v(a)} \int_a^b g\left(e^{is}\right) dv(s) \right| dt$$

$$\leq \frac{1}{2}|\Gamma - \gamma| L(b-a)$$

$$\times \max_{t \in [a,b]} \left| g\left(e^{it}\right) - \frac{1}{v(b) - v(a)} \int_a^b g\left(e^{is}\right) dv(s) \right|. \quad (7.51)$$

If $v \in \mathcal{M}^{\nearrow}[a,b]$, *then*

$$|G_{\exp}(f,g,v;a,b)|$$

$$\leq \frac{1}{2}|\Gamma - \gamma| \int_a^b \left| g\left(e^{it}\right) - \frac{1}{v(b) - v(a)} \int_a^b g\left(e^{is}\right) dv(s) \right| dv(t)$$

$$\leq \frac{1}{2}|\Gamma - \gamma| [v(b) - v(a)]$$

$$\times \max_{t \in [a,b]} \left| g\left(e^{it}\right) - \frac{1}{v(b) - v(a)} \int_a^b g\left(e^{is}\right) dv(s) \right|. \quad (7.52)$$

Proof. By Theorem 7.5, on replacing $g(e^{i\cdot})$ with $g(e^{i\cdot}) -$

$\frac{1}{v(b)-v(a)} \int_a^b g\left(e^{is}\right) dv\left(s\right)$ we get

$$\left| \int_a^b f\left(e^{it}\right) \left[g\left(e^{it}\right) - \frac{1}{v\left(b\right) - v\left(a\right)} \int_a^b g\left(e^{is}\right) dv\left(s\right) \right] dv\left(t\right) \right.$$

$$\left. - \frac{\gamma + \Gamma}{2} \int_a^b \left[g\left(e^{it}\right) - \frac{1}{v\left(b\right) - v\left(a\right)} \int_a^b g\left(e^{is}\right) dv\left(s\right) \right] dv\left(t\right) \right|$$

$$\leq \frac{1}{2} \left| \Gamma - \gamma \right| \int_a^b \left| g\left(e^{it}\right) - \frac{1}{v\left(b\right) - v\left(a\right)} \int_a^b g\left(e^{is}\right) dv\left(s\right) \right|$$

$$d\left(\bigvee_a^t \left(v\right) \right).$$

Since

$$\int_a^b f\left(e^{it}\right) \left[g\left(e^{it}\right) - \frac{1}{v\left(b\right) - v\left(a\right)} \int_a^b g\left(e^{is}\right) dv\left(s\right) \right] dv\left(t\right)$$

$$= \int_a^b f\left(e^{it}\right) g\left(e^{it}\right) dv\left(t\right) - \frac{1}{v\left(b\right) - v\left(a\right)} \int_a^b f\left(e^{it}\right) dv\left(t\right)$$

$$\int_a^b g\left(e^{it}\right) dv\left(t\right)$$

and

$$\int_a^b \left[g\left(e^{it}\right) - \frac{1}{v\left(b\right) - v\left(a\right)} \int_a^b g\left(e^{is}\right) dv\left(s\right) \right] dv\left(t\right) = 0,$$

hence the first inequality (7.50) is obtained. The second inequality is obvious.

The rest follow in a similar way and we omit the details.

Remark 32 *If $g\left(e^{i\cdot}\right)$ is of K-Lipschitzian type and v is of bounded variation, then [9]*

$$\left| g\left(e^{it}\right) \left[v\left(b\right) - v\left(a\right) \right] - \int_a^b g\left(e^{is}\right) dv\left(s\right) \right|$$

$$\leq K \left[\frac{1}{2} \left(b - a\right) + \left| t - \frac{a + b}{2} \right| \right] \bigvee_a^b \left(v\right),$$

for any $t \in [a, b]$.

  By (7.50) we then have

$$|G_{\exp}(f, g, v; a, b)|$$

$$\leq \frac{1}{2} \frac{|\Gamma - \gamma|}{|v(b) - v(a)|}$$

$$\int_a^b \left| g\left(e^{it}\right)[v(b) - v(a)] - \int_a^b g\left(e^{is}\right) dv(s) \right| d\left(\bigvee_a^t (v)\right)$$

$$\leq \frac{1}{2} \frac{|\Gamma - \gamma|}{|v(b) - v(a)|} K$$

$$\times \bigvee_a^b (v) \int_a^b \left| \frac{1}{2}(b-a) + \left| t - \frac{a+b}{2} \right| \right| d\left(\bigvee_a^t (v)\right).$$

Since, as above

$$\int_a^b \left[\frac{1}{2}(b-a) + \left| t - \frac{a+b}{2} \right| \right] d\left(\bigvee_a^t (v)\right)$$

$$= (b-a)\bigvee_a^b (v) - \int_a^b \operatorname{sgn}\left(t - \frac{a+b}{2}\right) \bigvee_a^t (v)\, dt$$

$$\leq (b-a)\bigvee_a^b (v),$$

then we get the following upper bounds for the magnitude of $G_{\exp}(f, g, v; a, b)$

$$|G_{\exp}(f, g, v; a, b)|$$

$$\leq \frac{1}{2} \frac{|\Gamma - \gamma|}{|v(b) - v(a)|} K$$

$$\times \bigvee_a^b (v) \left[(b-a)\bigvee_a^b (v) - \int_a^b \operatorname{sgn}\left(t - \frac{a+b}{2}\right) \bigvee_a^t (v)\, dt \right]$$

$$\leq \frac{1}{2} K \frac{|\Gamma - \gamma|(b-a)}{|v(b) - v(a)|} \left(\bigvee_a^b (v)\right)^2. \quad (7.53)$$

7.5 INEQUALITIES FOR UNITARY OPERATORS

For given x, $y \in H$, define the function $u(\lambda) := \langle E_\lambda x, y \rangle$, $\lambda \in [0, 2\pi]$. We know that u is of bounded variation and

$$\bigvee_0^{2\pi}(u) =: \bigvee_0^{2\pi}\left(\left\langle E_{(\cdot)}x, y\right\rangle\right) \le \|x\|\,\|y\|. \tag{7.54}$$

Proposition 7.10 (Dragomir 2019, [23]) *Let f, $g \in \mathcal{C}$ $(0, 1)$ and γ, $\Gamma \in \mathbb{C}$, $\gamma \ne \Gamma$ such that $f \in \bar{\Delta}_{\exp[0,2\pi]}(\gamma, \Gamma)$. Assume that U is a unitary operator on the Hilbert space $(H, \langle \cdot, \cdot \rangle)$ and $\{E_\lambda\}_{\lambda \in [0,2\pi]}$ the spectral family of U, then*

$$\left| \langle f(U)\, g(U)\, x, y \rangle - \frac{\gamma + \Gamma}{2} \langle g(U)\, x, y \rangle \right|$$

$$\le \frac{1}{2} |\Gamma - \gamma| \max_{z \in \mathcal{C}(0,1)} |g(z)| \bigvee_0^{2\pi}\left(\left\langle E_{(\cdot)}x, y\right\rangle\right)$$

$$\le \frac{1}{2} |\Gamma - \gamma| \max_{z \in \mathcal{C}(0,1)} |g(z)|\, \|x\|\,\|y\| \tag{7.55}$$

for all x, $y \in H$.
Moreover, if $g \in \bar{\Delta}_{\exp[0,2\pi]}(\delta, \Delta)$, then

$$\left| \langle f(U)\, g(U)\, x, y \rangle - \frac{\gamma + \Gamma}{2} \langle g(U)\, x, y \rangle \right.$$

$$\left. - \frac{\delta + \Delta}{2} \langle f(U)\, x, y \rangle + \frac{\gamma + \Gamma}{2} \cdot \frac{\delta + \Delta}{2} \langle x, y \rangle \right|$$

$$\le \frac{1}{4} |\Gamma - \gamma|\, |\Delta - \delta| \bigvee_0^{2\pi}\left(\left\langle E_{(\cdot)}x, y\right\rangle\right)$$

$$\le \frac{1}{4} |\Gamma - \gamma|\, |\Delta - \delta|\, \|x\|\,\|y\| \tag{7.56}$$

for all x, $y \in H$.

The proof follows by the inequalities (7.24), (7.35) and the spectral representation theorem.

Proposition 7.11 (Dragomir 2019, [23]) *Let* f, $g \in \mathcal{C}$ $(0,1)$ *and* γ, $\Gamma \in \mathbb{C}$, $\gamma \neq \Gamma$ *such that* $f \in \bar{\Delta}_{\exp[0,2\pi]}(\gamma,\Gamma)$. *Assume that* U *is a unitary operator on the Hilbert space* $(H, \langle \cdot, \cdot \rangle)$ *and* $\{E_\lambda\}_{\lambda \in [0,2\pi]}$ *the spectral family of* U *and* $g(e^{i\cdot}) \in \mathcal{BV}_{\mathbb{C}}[0,2\pi]$, *then*

$$\left| \langle f(U) g(U) x, y \rangle - \frac{\gamma + \Gamma}{2} \langle g(U) x, y \rangle \right.$$
$$\left. - g(1) \langle f(U) x, y \rangle + \frac{\gamma + \Gamma}{2} \cdot g(1) \langle x, y \rangle \right|$$
$$\leq \frac{1}{4} |\Gamma - \gamma| \bigvee_0^{2\pi} \left(g\left(e^i\right)\right) \bigvee_0^{2\pi} \left(\left\langle E_{(\cdot)} x, y \right\rangle \right)$$
$$\leq \frac{1}{4} |\Gamma - \gamma| \bigvee_0^{2\pi} \left(g\left(e^{i\cdot}\right)\right) \|x\| \|y\| \quad (7.57)$$

for all x, $y \in H$.

We also have

$$\left| \langle f(U) g(U) x, y \rangle - \frac{\gamma + \Gamma}{2} \langle g(U) x, y \rangle \right.$$
$$\left. - \langle f(U) x, y \rangle \frac{1}{2\pi} \int_0^{2\pi} g\left(e^{it}\right) dt + \langle x, y \rangle \frac{\gamma + \Gamma}{2} \cdot \right.$$
$$\left. \frac{1}{2\pi} \int_0^{2\pi} g\left(e^{it}\right) dt \right|$$
$$\leq \frac{1}{2} |\Gamma - \gamma| \max_{t \in [0,2\pi]} \left| g\left(e^{it}\right) - \frac{1}{2\pi} \int_0^{2\pi} g\left(e^{is}\right) ds \right|$$
$$\bigvee_0^{2\pi} \left(\left\langle E_{(\cdot)} x, y \right\rangle \right)$$
$$\leq \frac{1}{2} |\Gamma - \gamma| \max_{t \in [0,2\pi]} \left| g\left(e^{it}\right) - \frac{1}{2\pi} \int_0^{2\pi} g\left(e^{is}\right) ds \right| \|x\| \|y\|$$
$$(7.58)$$

for all x, $y \in H$.

The proof follows by the inequalities (7.36), (7.39) and the spectral representation theorem.

Proposition 7.12 (Dragomir 2019, [23]) *Let* f, $g \in \mathcal{C}$ $(0,1)$ *and* γ, $\Gamma \in \mathbb{C}$, $\gamma \neq \Gamma$ *such that* $f \in \bar{\Delta}_{\exp[0,2\pi]}(\gamma, \Gamma)$. *Assume that* U *is a unitary operator on the Hilbert space* $(H, \langle \cdot, \cdot \rangle)$ *and* $\{E_\lambda\}_{\lambda \in [0,2\pi]}$ *the spectral family of* U, *then*

$$|\langle x, y \rangle \langle f(U) g(U) x, y \rangle - \langle f(U) x, y \rangle \langle g(U) x, y \rangle|$$

$$\leq \frac{1}{2} |\Gamma - \gamma| \max_{t \in [0,2\pi]} \left| g\left(e^{it}\right) \langle x, y \rangle - \langle g(U) x, y \rangle \right| \bigvee_0^{2\pi} \left(\left\langle E_{(\cdot)} x, y \right\rangle \right)$$

$$\leq \frac{1}{2} |\Gamma - \gamma| \max_{t \in [0,2\pi]} \left| g\left(e^{it}\right) \langle x, y \rangle - \langle g(U) x, y \rangle \right| \|x\| \|y\|$$

$$\tag{7.59}$$

for all x, $y \in H$.

Moreover, if $g\left(e^{i\cdot}\right)$ *is of* K-*Lipschitzian type on* $[0, 2\pi]$, *then*

$$|\langle x, y \rangle \langle f(U) g(U) x, y \rangle - \langle f(U) x, y \rangle \langle g(U) x, y \rangle|$$

$$\leq \pi K |\Gamma - \gamma| \left(\bigvee_0^{2\pi} \left(\left\langle E_{(\cdot)} x, y \right\rangle \right) \right)^2 \leq \pi K |\Gamma - \gamma| \|x\|^2 \|y\|^2$$

for all x, $y \in H$.

The proof follows by the inequalities (7.50), (7.53) and the spectral representation theorem.

Bibliography

[1] T. M. Apostol. *Mathematical Analysis, Second Edition.* Addison-Wesley Pub. Com, 1975.

[2] N. S. Barnett, W.-S. Cheung, S. S. Dragomir, and A. Sofo. Ostrowski and trapezoid type inequalities for the Stieltjes integral with Lipschitzian integrands or integrators. *Comput. Math. Appl.*, 57(2):195–201, 2009.

[3] P. Cerone, W.-S. Cheung, and S. S. Dragomir. On Ostrowski type inequalities for Stieltjes integrals with absolutely continuous integrands and integrators of bounded variation. *Comput. Math. Appl.*, 54(2):183–191, 2007.

[4] P. Cerone and S. S. Dragomir. New bounds for the three-point rule involving the Riemann-Stieltjes integral. In *Advances in Statistics, Combinatorics and Related Areas*, pages 53–62. World Sci. Publ., River Edge, NJ, 2002.

[5] P. Cerone and S. S. Dragomir. Approximation of the Stieltjes integral and applications in numerical integration. *Appl. Math.*, 51(1):37–47, 2006.

[6] W.-S. Cheung and S. S. Dragomir. Two Ostrowski type inequalities for the Stieltjes integral of monotonic functions. *Preprints RGMIA Res. Rep. Coll.*, 9(3):1–8, 2006.

[7] S. S. Dragomir. The Ostrowski integral inequality for mappings of bounded variation. *Bull. Austral. Math. Soc.*, 60(3):495–508., 1999.

[8] S. S. Dragomir. On the Ostrowski's inequality for Riemann-Stieltjes integral. *Korean J. Appl. Math.*, 7(1):477–485, 2000.

[9] S. S. Dragomir. On the Ostrowski inequality for Riemann-Stieltjes integral $\int_a^b f(t)\, du(t)$ where f is of Hölder type and u is of bounded variation and applications. *J. KSIAM*, 5(1):35–45, 2001.

[10] S. S. Dragomir. On the Ostrowski's integral inequality for mappings with bounded variation and applications. *Math. Ineq. Appl.*, 4(1):59–66, 2001.

[11] S. S. Dragomir. Some inequalities for Riemann-Stieltjes integral and applications. In *Optimization and Related Topics*, pages 197–235. Kluwer Academic Publishers, 2001.

[12] S. S. Dragomir. Approximating the Riemann–Stieltjes integral by a trapezoidal quadrature rule with applications. *Mathematical and Computer Modelling*, 54(1-2):243–260, 2002.

[13] S. S. Dragomir. Sharp bounds of Čebyšev functional for Stieltjes integrals and applications. *Bull. Austral. Math. Soc.*, 67(2):257–266, 2003.

[14] S. S. Dragomir. Inequalities of Grüss type for the Stieltjes integral and applications. *Kragujevac J. Math.*, 26(1):89–122, 2004.

[15] S. S. Dragomir. A generalisation of Cerone's identity and applications. *Tamsui Oxf. J. Math. Sci.*, 23(1):79–90, 2007.

[16] S. S. Dragomir. Inequalities for Stieltjes integrals with convex integrators and applications. *Appl. Math. Lett.*, 20(3):123–130, 2007.

[17] S. S. Dragomir. Generalised trapezoid-type inequalities for complex functions defined on unit circle with applications for unitary operators in Hilbert spaces. *Mediterr. J. Math.*, 12(3):573–591, 2015.

[18] S. S. Dragomir. Grüss type inequalities for complex functions defined on unit circle with applications for unitary operators in Hilbert spaces. *Rev. Colombiana Mat.*, 49(1):77–94, 2015.

[19] S. S. Dragomir. Ostrowski's type inequalities for complex functions defined on unit circle with applications for unitary operators in Hilbert spaces. *Arch. Math. (Brno)*, 51(4):233–254, 2015.

[20] S. S. Dragomir. Quasi Grüss type inequalities for complex functions defined on unit circle with applications for unitary operators in Hilbert spaces. *Extracta Math.*, 31(1):47–67, 2016.

[21] S. S. Dragomir. Trapezoid type inequalities for complex functions defined on the unit circle with applications for unitary operators in Hilbert spaces. *Georgian Math. J.*, 23(2):199–210, 2016.

[22] S. S. Dragomir. Ostrowski type inequalities for Lebesgue integral: a survey of recent results. *Aust. J. Math. Anal. Appl.*, 14(Art. 1):1–283, 2017.

[23] S. S. Dragomir. Some weighted inequalities for Riemann-Stieltjes integral when a function is bounded. *RGMIA Res. Rep. Coll.*, to appear:1–23, 2019.

[24] S. S. Dragomir, C. Buse, M. V. Boldea, and L. Braescu. A generalization of the trapezoidal rule for the Riemann-Stieltjes integral and applications. *Nonlinear Anal. Forum*, 6(2):337–351, 2001.

[25] S. S. Dragomir and I. A. Fedotov. An inequality of Grüss type for the Riemann-Stieltjes integral and applications for special means. *Tamkang J. Math.*, 29(4):287–292, 1998.

[26] S. S. Dragomir and I. A. Fedotov. A Grüss type inequality for mappings of bounded variation and applications for numerical analysis. *Nonlinear Funct. Anal. Appl.*, 6(3):425–433, 2001.

[27] G. Helmberg. *Introduction to Spectral Theory in Hilbert Space*. John Wiley, New York, NY, USA, 1969.

[28] A. Ostrowski. Über die absolutabweichung einer differentiierbaren funktion von ihrem integralmittelwert (German). *Comment. Math. Helv.*, 10(1):226–227, 1938.